A PORTABLE LATIN
for Gardeners

The University of Chicago Press, Chicago 60637

© 2016 Quid Publishing

All rights reserved. Published 2016.

Printed in China

Conceived, designed, and produced by

Part of the Quarto Group

Quid Publishing

Ovest House

58 West Street

Brighton

BN1 2RA

England

Original book design: Lindsey Johns

Layout design: Tony Seddon

26 25 24 23 22 21 20 19 18 17   1 2 3 4 5

ISBN-13: 978-0-226-45536-5   (paper)

ISBN-13: 978-0-226-45553-2   (e-book)

DOI: 10.7208/chicago/9780226455532.001.0001

Library of Congress Control Number: 2016941002

A PORTABLE LATIN
for Gardeners

More than 1,500 Essential Plant Names and the Secrets They Contain

---

JAMES ARMITAGE

THE UNIVERSITY
OF CHICAGO PRESS
CHICAGO

Contents

CHAPTER ONE
COLOR

CHAPTER TWO
PLANT FORM

CHAPTER THREE
FEATURES OF PLANTS

CHAPTER FOUR
COMPARISONS

CHAPTER FIVE
PLACES AND PEOPLE

CHAPTER SIX
IDEAS, ASSOCIATIONS, AND PROPERTIES

Introduction

Lagurus ovatus

The perception is common that gardening would be a lot more fun if we did not have to talk about plants using a dead language made up of unpronounceable words. It is hard not to concede the point that, to an English-speaker, a lot of Latin names don't exactly roll off the tongue, and that it is undoubtedly easier to remember the name dawn redwood than *Metasequoia glyptostroboides*. So why do we bother with it at all?

The first thing to understand is that we require plant names to be universal. Flip through a botanical or horticultural text in Korean or Mandarin and you may not understand a single word except the plant names. They, at least, are immune to the barriers of language. This is important for scientific communication, for conservation, and the sharing of understanding, but no less for gardening. The plants we grow in our gardens come from all over the world and if we want to gain reliable knowledge about them it is necessary we all use the same names. In this way the great advantage of Latin is the very fact that it is a dead language; it is static and anchored by unchanging rules, it is no one's and everyone's at the same time.

Having a specialized system for naming also allows us to group plants together in informative ways. There is nothing in the common names skunkbush, stag's-horn sumach, and wax tree to suggest any particular relationship or similarity. However, once you know all three are species of the genus *Rhus*, then immediately you can say each is likely to have pinnate leaves and good fall color.

This is, perhaps, fairly well understood, but what is often overlooked is that Latin plant names are like little parcels of information about the plants to which they refer. From flower color to flavor and habit to habitat, Latin names have a huge amount of information to impart that is of practical value to gardeners. Sometimes, the relevance is clear. Knowing that the name of *Rudbeckia laciniata* indicates that it has strongly cut leaves or that *Digitalis purpurea f. albiflora* was named for its milky white flowers is of obvious interest. However, being aware that *Griselinia littoralis* grows by the sea or *Convallaria majalis* flowers in May might be considered just as useful.

There are a host of names that tell us something of practical value. Demonstrating this, and making the information easily accessible, is the purpose of this book. As with other dictionaries of plant names, definitions are given but here, instead of providing a simple alphabetical listing, the names are first divided into categories that indicate the kinds of information they contain. This provides a fascinating view of the range and coverage of plant names and it is hoped will enhance an appreciation of the things they have to tell us. Once it is understood that Latin plant names are practical tools and not just an inconvenient necessity, they can help us enjoy the plants around us and add greatly to the pleasure of gardening.

Primula pulverulenta

How to use this book

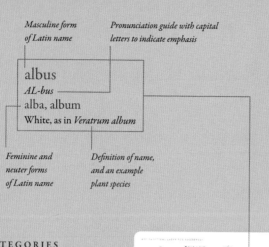

Masculine form of Latin name

Pronunciation guide with capital letters to indicate emphasis

albus
AL-bus
alba, album
White, as in *Veratrum album*

Feminine and neuter forms of Latin name

Definition of name, and an example plant species

ILLUSTRATIONS
Botanical watercolors are used to show specific plant features and demonstrate the meaning of key Latin words.

CATEGORIES
Latin words are grouped into different categories, such as "Bright colors," so that readers can instantly see which names share a common theme.

INDEX
Alphabetical listing allows readers to look up specific words and use the book as a conventional dictionary.

BEHIND THE NAME
Feature boxes reveal the hidden knowledge that can be gained from a plant's Latin name, along with other information of interest.

A brief guide to plant names

In writing plant names certain conventions should be followed.

Family
(e.g. *Magnoliaceae*)

Family names end in *–aceae* and should appear in italics with an upper-case initial letter.

Genus (plural genera)
(e.g. *Magnolia*)

This appears in italics with an upper-case initial letter. There are three genders: masculine, feminine, and neuter. The genus may be abbreviated to its initial letter to avoid repetition, e.g. *Magnolia kobus, M. stellata.*

Species
(e.g. *Magnolia grandiflora*)

The species is a specific unit within a genus and the name is sometimes referred to as the specific epithet. It is written in lower-case italics. The combination of the genus and species name gives us the binomial, or two-word, system which is the basis of modern nomenclature. Most species names are adjectives and need to agree with the gender of the genus.

Subspecies, varietas and forma
(e.g. *Magnolia sprengeri* var. *diva*)

There are three ranks commonly found within species, these are known as infraspecific names. All appear in italics and are preceded by an abbreviation of their rank in roman. Subspecies (subsp.) is a distinct variant of the species; varietas (var.), sometimes known as variety, distinguishes slight variations; forma (f.), sometimes known as form, also distinguishes minor variations.

Cultivar
(e.g. *Magnolia sieboldii* 'Colossus')

This appears as lower-case roman type with an upper-case initial letter within single quotes. It is applied to artificially maintained plants.

Hybrid
(e.g. *Magnolia* × *soulangeana*)

This appears as lower-case italics and is preceded by a roman-type multiplication sign. It is applied to plants that are the product of a cross between species of the same genus.

Poppy anemone,
Anemone coronaria

This is a simplified outline of the binomial system but omits many complexities. The purpose of this book is to help you to understand what names mean, not how they are formulated. However, for those with an interest in the intricacies of botanical nomenclature, online resources are available.

CHAPTER ONE

Color

Color is all-important to gardeners and the primary means by which mood is evoked in planting schemes. Imagine passing along the Flower Garden Walk at Longwood Gardens, Pennsylvania, or admiring the sweeps of color in the Butchart Gardens, British Columbia, and the impact of color in the garden becomes clear. There are a wide selection of names that refer to color and these can offer helpful hints to the horticulturist trying to hit just the right note.

Light colors

Light colors help to give a sense of openness to the garden and are therefore useful in confined spaces or shady sites. They are subtle, gentle, calming, and the perfect visual accompaniment to the sounds of birdsong and running water. Names meaning silver or gray often refer not to plants' flowers but to their leaves. It is worth knowing that such plants generally enjoy a bright spot and free-draining soil.

albescens
al-BES-enz
Becoming white, as in *Kniphofia albescens*

albicans
AL-bih-kanz
Off-white, as in *Hebe albicans*

albidus
AL-bi-dus
albida, albidum
White, as in *Trillium albidum*

albiflorus
al-BIH-flor-us
albiflora, albiflorum
With white flowers, as in *Buddleja albiflora*

albifrons
AL-by-fronz
With white fronds, as in *Cyathea albifrons*

albomarginatus
AL-bow-mar-gin-AH-tus
albomarginata, albomarginatum
With white margins, as in *Agave albomarginata*

albosinensis
al-bo-sy-NEN-sis
albosinensis, albosinense
Meaning white and from China, as in *Betula albosinensis*

◀ The cemetery iris, *Iris albicans*, is native to Saudi Arabia but is commonly planted by gravesides in other places with a warm, dry climate. The Latin name refers to the off-white color of the flowers.

albus
AL-bus
alba, album
White, as in *Veratrum album*

argentatus
ar-jen-TAH-tus
argentata, argentatum

argenteus
ar-JEN-tee-us
argentea, argenteum
Silver in color, as in *Salvia argentea*

argyrophyllus
ar-ger-o-FIL-us
argyrophylla, argyrophyllum
With silver leaves, as in *Rhododendron argyrophyllum*

candicans
KAN-dee-kanz

candidus
KAN-dee-dus
candida, candidum
Shining white, as in *Echium candicans*

canescens
kan-ESS-kenz
With off-white or gray hairs, as in *Populus* × *canescens*

chionanthus
kee-on-AN-thus
chionantha, chionanthum
With snow-white flowers, as in *Primula chionantha*

chloropetalus
klo-ro-PET-al-lus
chloropetala,
chloropetalum
With green petals, as in *Trillium chloropetalum*

chrysanthus
kris-AN-thus
chrysantha, chrysanthum
With golden flowers, as in *Crocus chrysanthus*

chrysocarpus
kris-oh-KAR-pus
chrysocarpa,
chrysocarpum
With golden fruit, as in *Crataegus chrysocarpa*

chrysolepis
kris-SOL-ep-is
chrysolepis, chrysolepe
With golden scales, as in *Quercus chrysolepis*

chrysophyllus
kris-oh-FIL-us
chrysophylla,
chrysophyllum
With golden leaves, as in *Phlomis chrysophylla*

cinerarius
sin-uh-RAH-ree-us
cineraria, cinerarium
Ash-gray, as in *Centaurea cineraria*

cinerascens
sin-er-ASS-enz
Turning to ash-gray, as in *Senecio cinerascens*

The fragrant Madonna lily, *Lilium candidum*, is a symbol of purity and is often depicted in association with the Virgin Mary. The epithet *candidum* means shining white, an apt reference to its beautiful flowers.

cinereus
sin-EER-ee-us
cinerea, cinereum
The color of ash, as in *Veronica cinerea*

cinnamomeus
sin-uh-MOH-mee-us
cinnamomea,
cinnamomeum
Cinnamon-brown, as in *Osmunda cinnamomea*

citrinus
sit-REE-nus
citrina, citrinum
Lemon-yellow or like *Citrus*, as in *Callistemon citrinus*

coelestinus
koh-el-es-TEE-nus
coelestina, coelestinum

coelestis
koh-el-ES-tis
coelestis, coeleste
Sky-blue, as in *Phalocallis coelestis*

eburneus
eb-URN-ee-us
eburnea, eburneum
Ivory-white, as in *Angraecum eburneum*

flavens
flav-ENZ

flaveolus
fla-VEE-oh-lus
flaveola, flaveolum

flavescens
flav-ES-enz

flavidus
FLA-vid-us
flavida, flavidum
Various kinds of yellow, as in *Anigozanthos flavidus*

flavicomus
flay-vih-KOH-mus
flavicoma, flavicomum
With yellow hairs, as in *Euphorbia flavicoma*

flavovirens
fla-voh-VY-renz
Greenish yellow, as in *Callistemon flavovirens*

flavus
FLA-vus
flava, flavum
Pure yellow, as in *Crocus flavus*

fulvidus
FUL-vee-dus
fulvida, fulvidum
Slightly tawny in color, as in
Cortaderia fulvida

fulvus
FUL-vus
fulva, fulvum
Tawny orange in color, as in
Hemerocallis fulva

gilvus
GIL-vus
gilva, gilvum
Dull yellow, as in *Echeveria ×
gilva*

glaucescens
glaw-KES-enz
With a bloom (a powdery
deposit); blue-green in color,
as in *Ferocactus glaucescens*

glaucophyllus
glaw-koh-FIL-us
glaucophylla,
glaucophyllum
With gray-green leaves, or with a
bloom (a powdery deposit), as in
Rhododendron glaucophyllum

◧ The blue-leaved rhodo-
dendron, *Rhododendron
glaucophyllum,* has pinkish
flowers. The name
glaucophyllum, meaning with
gray-green or bloomed leaves,
refers to the pale underside of
the foliage.

glaucus
GLAW-kus
glauca, glaucum
With a bloom on the leaves, as in
Festuca glauca

griseus
GREE-see-us
grisea, griseum
Gray, as in *Acer griseum*

holochrysus
ho-loh-KRIS-us
holochrysa, holochrysum
Completely golden, as in
Aeonium holochrysum

hypoleucus
hy-poh-LOO-kus
hypoleuca, hypoleucum
White underneath, as in
Centaurea hypoleuca

incanus
in-KAN-nus
incana, incanum
Gray, as in *Geranium incanum*

incarnatus
in-kar-NAH-tus
incarnata, incarnatum
The color of flesh, as in
Dactylorhiza incarnata

lacteus
lak-TEE-us
Milk-white, as in *Cotoneaster
lacteus*

lactiflorus
lak-tee-FLOR-us
lactiflora, lactiflorum
With milk-white flowers, as in
Campanula lactiflora

laetevirens
lay-tee-VY-renz
Vivid green, as in *Parthenocissus
laetevirens*

leuc-
Used in compound words to
denote white

leucochilus
loo-KOH-ky-lus
leucochila, leucochilum
With white lips, as in *Oncidium
leucochilum*

leucodermis
loo-koh-DER-mis
leucodermis, leucoderme
With white skin, as in *Rubus
leucodermis*

leuconeurus
loo-koh-NOOR-us
leuconeura, leuconeurum
With white nerves, as in
Maranta leuconeura

leucophaeus
loo-koh-FAY-us
leucophaea, leucophaeum
Dusky white, as in *Dianthus leucophaeus*

lilacinus
ly-luc-SEE-nus
lilacina, lilacinum
Lilac, as in *Primula lilacina*

lividus
LI-vid-us
livida, lividum
Blue-gray; the color of lead, as in *Helleborus lividus*

luridus
LEW-rid-us
lurida, luridum
Pale yellow, wan, as in *Moraea lurida*

luteolus
loo-tee-OH-lus
luteola, luteolum
Yellowish, as in *Primula luteola*

malvinus
mal-VY-nus
malvina, malvinum
Mauve, as in *Plectranthus malvinus*

nivalis
niv-VAH-lis
nivalis, nivale

niveus
NIV-ee-us
nivea, niveum

nivosus
niv-OH-sus
nivosa, nivosum
As white as snow, or growing near snow, as in *Galanthus nivalis*

ochroleucus
ock-roh-LEW-kus
ochroleuca, ochroleucum
Yellowish white, as in *Crocus ochroleucus*

phaeacanthus
fay-uh-KAN-thus
phaeacantha, phaeacanthum
With gray thorns, as in *Opuntia phaeacantha*

polifolius
po-lih-FOH-lee-us
polifolia, polifolium
With gray leaves, as in *Andromeda polifolia*

senescens
sen-ESS-enz
Seeming to grow old (i.e. white or gray), as in *Allium senescens*

sulphureus
sul-FER-ee-us
sulphurea, sulphureum
Sulfur-yellow, as in *Lilium sulphureum*

xanthinus
zan-TEE-nus
xanthina, xanthinum
Yellow, as in *Rosa xanthina*

xanthocarpus
zan-tho-KAR-pus
xanthocarpa, xanthocarpum
With yellow fruits, as in *Rubus xanthocarpus*

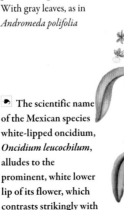

🢂 The scientific name of the Mexican species white-lipped oncidium, *Oncidium leucochilum*, alludes to the prominent, white lower lip of its flower, which contrasts strikingly with the rest of the bloom.

Bright colors

Bright colors lift the senses more immediately than almost anything else. In hanging baskets, bedding, and borders it is entirely permissible to give in to the unbridled vivacity of rich colors. Nature does not recognize the concept of clashing colors and it can be a source of great joy to bring this lesson into the garden. However, where a particular color scheme is desired, plant names can give some useful guidance.

amethystinus
am-eth-ih-STEE-nus
amethystina, amethystinum
Violet, as in *Brimeura amethystina*

aurantiacus
aw-ran-ti-AH-kus
aurantiaca, aurantiacum

aurantius
aw-RAN-tee-us
aurantia, aurantium
Orange, as in *Pilosella aurantiaca*

auratus
aw-RAH-tus
aurata, auratum
With golden rays, as in *Lilium auratum*

aureosulcatus
aw-ree-oh-sul-KAH-tus
aureosulcata, aureosulcatum
With yellow furrows, as in *Phyllostachys aureosulcata*

aureus
AW-re-us
aurea, aureum
Golden yellow, as in *Phyllostachys aurea*

auricomus
aw-RIK-oh-mus
auricoma, auricomum
With golden hair, as in *Ranunculus auricomus*

The cardinal flower, *Lobelia cardinalis*, lives up to its name with its crimson flowers but in several cultivars these are accompanied by bronze foliage. Being a plant of damp places it will not thrive in dry soils.

azureus
a-ZOOR-ee-us
azurea, azureum
Azure; sky-blue, as in *Muscari azureum*

caerulescens
see-roo-LES-enz
Turning blue, as in *Euphorbia caerulescens*

caesius
KESS-ee-us
caesia, caesium
Blue-gray, as in *Allium caesium*

cardinalis
kar-dih-NAH-lis
cardinalis, cardinale
Bright scarlet; cardinal-red, as in *Lobelia cardinalis*

carminatus
kar-MIN-uh-tus
carminata, carminatum

carmineus
kar-MIN-ee-us
carminea, carmineum
Carmine; bright crimson, as in *Metrosideros carminea*

carneus
KAR-nee-us
carnea, carneum
Flesh color; deep pink, as in *Androsace carnea*

cinnabarinus
sin-uh-bar-EE-nus
cinnabarina,
cinnabarinum
Vermilion, as in *Echinopsis cinnabarina*

coccineus
kok-SIN-ee-us
coccinea, coccineum
Scarlet, as in *Musa coccinea*

coeruleus
ko-er-OO-lee-us
coerulea, coeruleum
Blue, as in *Satureja coerulea*

coralliflorus
kaw-lih-FLOR-us
coralliflora, coralliflorum
With coral-red flowers, as in *Lampranthus coralliflorus*

corallinus
kor-al-LEE-nus
corallina, corallinum
Coral-red, as in *Ilex corallina*

crocatus
kroh-KAH-tus
crocata, crocatum

croceus
KRO-kee-us
crocea, croceum
Saffron-yellow, as in *Tritonia crocata*

cruentus
kroo-EN-tus
cruenta, cruentum
Bloody, as in *Lycaste cruenta*

🌺 Bachelor's buttons, *Centaurea cyanus*, is not strictly speaking native to the UK but is an ancient agricultural weed. Once common, it is now scarce but its blue flowers make it a pretty plant for the cottage garden.

cyananthus
sy-an-NAN-thus
cyanantha, cyananthum
With blue flowers, as in *Penstemon cyananthus*

cyaneus
sy-AN-ee-us
cyanea, cyaneum

cyanus
sy-AH-nus
Blue, as in *Allium cyaneum*

erubescens
er-oo-BESS-enz
Becoming red; blushing, as in *Philodendron erubescens*

erythro-
Used in compound words to denote red

euchlorus
YOO-klor-us
euchlora, euchlorum
A healthy green, as in *Tilia* × *euchlora*

flammeus
FLAM-ee-us
flammea, flammeum
Flame-colored; flame-like, as in *Tigridia flammea*

fucatus
few-KAH-tus
fucata, fucatum
Painted; dyed, as in *Crocosmia fucata*

ignescens
ig-NES-enz

igneus
ig-NE-us
ignea, igneum
Fiery red, as in *Cuphea ignea*

ionanthus
eye-oh-NAN-thus
ionantha, ionanthum
With violet-colored flowers, as in *Saintpaulia ionantha*

iridescens
ir-id-ES-enz
Iridescent, as in *Phyllostachys iridescens*

◄ **It is for its highly ornamental, violet-colored flowers that the tropical lilac tree, *Lonchocarpus violaceus*, received the epithet *violaceus*. But beware: this tender plant is as poisonous as it is pretty.**

lateritius
la-ter-ee-TEE-us
lateritia, lateritium
Brick-red, as in *Kalanchoe lateritia*

miniatus
min-ee-AH-tus
miniata, miniatum
Cinnabar-red, as in *Clivia miniata*

pavoninus
pav-ON-ee-nus
pavonina, pavoninum
Peacock-blue, as in *Anemone pavonina*

pictus
PIK-tus
picta, pictum
Painted; highly colored, as in *Acer pictum*

rhodanthus
rho-DAN-thus
rhodantha, rhodanthum
With rose-colored flowers, as in *Mammillaria rhodantha*

roseus
RO-zee-us
rosea, roseum
Colored like rose (*Rosa*), as in *Lapageria rosea*

rubellus
roo-BELL-us
rubella, rubellum
Pale red, becoming red, as in *Peperomia rubella*

rubens
ROO-benz

ruber
ROO-ber
rubra, rubrum
Red, as in *Plumeria rubra*

rubescens
roo-BES-enz
Becoming red, as in *Salvia rubescens*

rubricaulis
roo-bri-KAW-lis
rubricaulis, rubricaule
With red stems, as in *Actinidia rubricaulis*

rufinervis
roo-fi-NER-vis
rufinervis, rufinerve
With red veins, as in *Acer rufinerve*

rufus
ROO-fus
rufa, rufum
Red, as in *Prunus rufa*

testaceus
test-AY-see-us
testacea, testaceum
Brick-colored, as in *Lilium × testaceum*

violaceus
vy-oh-LAH-see-us
violacea, violaceum
Violet, as in *Hardenbergia violacea*

virens
VEER-enz
Green, as in *Penstemon virens*

virescens
veer-ES-enz
Turning green, as in *Carpobrotus virescens*

viridescens
vir-ih-DESS-enz
Turning green, as in *Ferocactus viridescens*

viridiflorus
vir-id-uh-FLOR-us
viridiflora, viridiflorum
With green flowers, as in *Lachenalia viridiflora*

viridis
VEER-ih-dis
viridis, viride
Green, as in *Trillium viride*

viridissimus
vir-id-ISS-ih-mus
viridissima, viridissimum
Very green, as in *Forsythia viridissima*

viridistriatus
vi-rid-ee-stry-AH-tus
viridistriata, viridistriatum
With green stripes, as in
Pleioblastus viridistriatus

viridulus
vir-ID-yoo-lus
viridula, viridulum
Rather green, as in *Tricyrtis
viridula*

vitellinus
vy-tel-LY-nus
vitellina, vitellinum
The color of egg yolk, as in
Encyclia vitellina

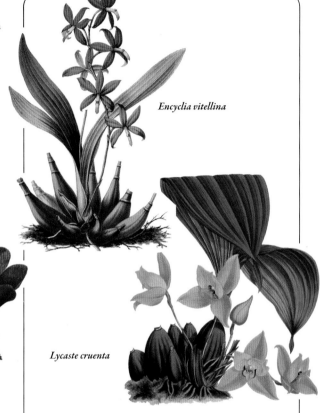

Encyclia vitellina

Lycaste cruenta

◖ The windflower, *Anemone
pavonina* (right), is a
tuberous perennial from
the Mediterranean. In addition
to the purplish peacock-colored
flowers suggested by its name,
this species may bear red or
pink blooms.

BEHIND THE NAME
Cryptic orchid names

Plant names can seem cryptic. The epithet
vitellina means "the color of egg yolk" but flowers
of *Encyclia vitellina* are generally reddish with only
a small central yellow part. It is difficult initially
to see how *Lycaste cruenta*, the blood-stained
maxillaria, earned either its common or scientific
name, which means bloody. Closer inspection
reveals the red patch on the lip.

Dark colors and multicolors

When used in excess, dark colors can seem brooding, oppressive, and heavy. However, they add depth and substance in small species and perspective in large ones. Dark colors are invaluable as a foil to fiery bedding and, by the strength of their contrast, lend luminosity to paler tones. Understanding how to use deep shades can add drama to garden compositions. Terms relating to more than one color are also included here.

androsaemus
an-dro-SAY-mus
androsaema,
androsaemum
With sap the color of blood, as in *Hypericum androsaemum*

atrocarpus
at-ro-KAR-pus
atrocarpa, atrocarpum
With black or very dark fruit, as in *Berberis atrocarpa*

atropurpureus
at-ro-pur-PURR-ee-us
atropurpurea,
atropurpureum
Dark purple, as in *Scabiosa atropurpurea*

atrorubens
at-roh-ROO-benz
Dark red, as in *Helleborus atrorubens*

atrosanguineus
at-ro-san-GWIN-ee-us
atrosanguinea,
atrosanguineum
Dark blood red, as in *Rhodochiton atrosanguineus*

atrovirens
at-ro-VY-renz
Dark green, as in *Chamaedorea atrovirens*

bicolor
BY-kul-ur
With two colors, as in *Caladium bicolor*

brunneus
BROO-nee-us
brunnea, brunneum
Deep brown, as in *Coprosma brunnea*

cupreatus
kew-pree-AH-tus
cupreata, cupreatum

cupreus
kew-pree-US
cuprea, cupreum
The color of copper, as in *Alocasia cuprea*

discolor
DIS-kol-or
Of two quite different colors, as in *Salvia discolor*

📷 Sun roses, *Helianthemum*, are plants of open, dry places that make excellent specimens for the rock garden. Their large, tissue-paper petals are often red or yellow or, as in this species, *H. cupreum*, copper-colored.

mediopictus
MED-ee-o-pic-tus
mediopicta, mediopictum
With a stripe or color running
down the middle, as in *Calathea
mediopicta*

melanocaulon
mel-an-oh-KAW-lon
With black stems, as in
Blechnum melanocaulon

melanoxylon
mel-an-oh-ZY-lon
With black wood, as in *Acacia
melanoxylon*

● The white, frothy flowers of
the elderberry, *Sambucus nigra*,
are a common sight of early
summer but it is not for these
that this species gained its
name, which means black, but
for its dark fruit in fall.

ferrugineus
fer-oo-GIN-ee-us
ferruginea, ferrugineum
Rust-colored, as in *Digitalis
ferruginea*

fuscus
FUS-kus
fusca, fuscum
Dusky or swarthy brown, as in
Nothofagus fusca

haematocalyx
hem-at-oh-KAL-icks
With a blood-red calyx, as in
Dianthus haematocalyx

haematochilus
hem-mat-oh-KY-lus
haematochila,
haematochilum
With a blood-red lip, as in
Oncidium haematochilum

haematodes
hem-uh-TOH-deez
Blood-red, as in *Rhododendron
haematodes*

hyacinthinus
hy-uh-sin-THEE-nus
hyacinthina,
hyacinthinum
hyacinthus
hy-uh-SIN-thus
hyacintha, hyacinthum
Dark purple-blue, or like a
hyacinth, as in *Triteleia
hyacinthina*

hypochondriacus
hy-po-kon-dree-AH-kus
hypochondriaca,
hypochondriacum
With a melancholy appearance;
with dull-colored flowers, as in
Amaranthus hypochondriacus

kermesinus
ker-mes-SEE-nus
kermesina, kermesinum
Crimson, as in *Passiflora
kermesina*

niger
NY-ger
nigra, nigrum
Black, as in *Phyllostachys nigra*

nigrescens
ny-GRESS-enz
Turning black, as in *Silene
nigrescens*

nigricans
ny-GRIH-kanz
Blackish, as in *Kennedia
nigricans*

ochraceus
oh-KRA-see-us
ochracea, ochraceum
Ocher-colored, as in *Hebe
ochracea*

opacus
oh-PAH-kus
opaca, opacum
Dark; dull; shaded, as in
Crataegus opaca

 The bloody cranesbill, *Geranium sanguineum*, is a rugged little plant that also makes a splendid specimen for a border or rock garden. A variety, var. *striatum*, has pale pink flowers with crimson veins.

phaeus
FAY-us
phaea, phaeum
Dusky, as in *Geranium phaeum*

phoeniceus
fee-nik-KEE-us
phoenicea, phoeniceum
Purple-red, as in *Juniperus phoenicea*

phoenicolasius
fee-nik-oh-LASS-ee-us
phoenicolasia, phoenicolasium
With purple hairs, as in *Rubus phoenicolasius*

polychromus
pol-ee-KROW-mus
polychroma, polychromum
With many colors, as in *Masdevallia polychroma*

puniceus
pun-IK-ee-us
punicea, puniceum
Red-purple, as in *Clianthus puniceus*

purpurascens
pur-pur-ASS-kenz
Becoming purple, as in *Bergenia purpurascens*

purpuratus
pur-pur-AH-tus
purpurata, purpuratum
Made purple, as in *Phyllostachys purpurata*

purpureus
pur-PUR-ee-us
purpurea, purpureum
Purple, as in *Digitalis purpurea*

rubiginosus
roo-bij-ih-NOH-sus
rubiginosa, rubiginosum
Rusty, as in *Ficus rubiginosa*

sanguineus
san-GWIN-ee-us
sanguinea, sanguineum
Blood-red, as in *Geranium sanguineum*

stygianus
sty-jee-AH-nus
stygiana, stygianum
Dark, as in *Euphorbia stygiana*

tricolor
TRY-kull-lur
With three colors, as in *Tropaeolum tricolor*

versicolor
VER-suh-kuh-lor
With various colors, as in *Oxalis versicolor*

 Commonly known as pigsqueak, bergenias are perhaps primarily admired for their bold, glossy foliage, but elephant ears, *Bergenia purpurascens*, also bears richly purple flowers, alluded to in its name. They are highly ornamental.

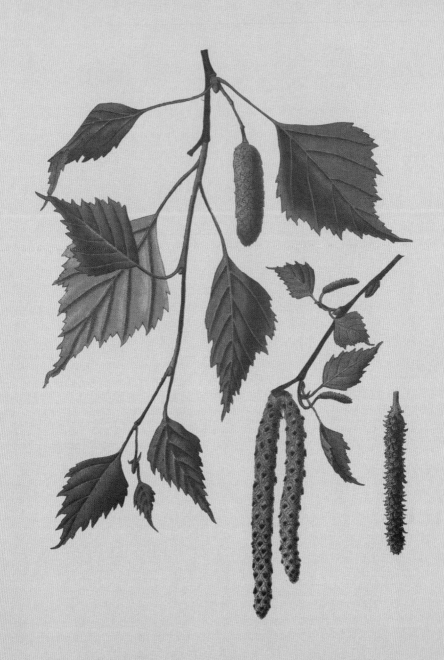

CHAPTER TWO

Plant Form

Correct positioning of plants is key to effective garden design. It is perfectly possible to ruin an otherwise balanced scheme by misplacing a single specimen. Names that give some sense of a plant's stature or structure, therefore, can provide clues important to successful gardening. Plant form does not always refer to overall appearance, however. Also in this chapter are names that refer to smaller characteristics, perhaps of the leaves, flowers, or fruit; those details that really help us know the plants we grow.

Habit

The habit a plant adopts gives it so much of its character. Picture the upright stance of the noble, reliable and dignified Irish yew, *Taxus baccata* 'Fastigiata', or the wild, free, romantic sweeping branches of the European white birch, *Betula pendula*. For gardeners, knowing a plant's growth habit can be very useful when it comes to choosing plants for the garden. In this regard, some names can help toward making an informed choice.

acaulis
a-KAW-lis
acaulis, acaule
Without a stem, or short-stemmed, as in *Gentiana acaulis*

adscendens
ad-SEN-denz
Ascending; rising, as in *Aster adscendens*

arborescens
ar-bo-RES-senz

arboreus
ar-BOR-ee-us
arborea, arboreum
A woody or tree-like plant, as in *Erica arborea*

ascendens
as-SEN-denz
Rising upward, as in *Calamintha ascendens*

bifurcatus
by-fur-KAH-tus
bifurcata, bifurcatum
Divided into equal stems or branches, as in *Platycerium bifurcatum*

The stemless gentian, *Gentiana acaulis*, owes both its common and scientific names to its short flowering stems that help accentuate the effect of its upward-facing blooms.

caespitosus
kess-pi-TOH-sus
caespitosa, caespitosum
Growing in a dense clump, as in *Eschscholzia caespitosa*

catarractae
kat-uh-RAK-tay
Of waterfalls, as in *Parahebe catarractae*

caulescens
kawl-ESS-kenz
With a stem, as in *Kniphofia caulescens*

cernuus
SER-new-us
cernua, cernuum
Drooping or nodding, as in *Enkianthus cernuus*

columnaris
kol-um-NAH-ris
columnaris, columnare
In the shape of a column, as in *Eryngium columnare*

comans
KO-manz

comatus
kom-MAH-tus
comata, comatum
Tufted, as in *Carex comans*

comosus
kom-OH-sus
comosa, comosum
With tufts, as in *Eucomis comosa*

compactus
kom-PAK-tus
compacta, compactum
Compact; dense, as in *Pleiospilos compactus*

complexus
kom-PLEKS-us
complexa, complexum
Complex; encircled; as in *Muehlenbeckia complexa*

condensatus
kon-den-SAH-tus
condensata, condensatum

condensus
kon-DEN-sus
condensa, condensum
Crowded together, as in *Alyssum condensatum*

congestus
kon-JES-tus
congesta, congestum
Congested or crowded together, as in *Aciphylla congesta*

conglomeratus
kon-glom-er-AH-tus
conglomerata, conglomeratum
Crowded together, as in *Cyperus conglomeratus*

contortus
kon-TOR-tus
contorta, contortum
Twisted; contorted, as in *Pinus contorta*

contractus
kon-TRAK-tus
contracta, contractum
Contracted; drawn together, as in *Fargesia contracta*

◖ The split rock plant, *Pleiospilos compactus*, is a remarkable species with paired, succulent leaves resembling stones cleft in two. In combination with its very compact habit, this camouflages the plant in the stony places where it grows.

crassus
KRASS-us
crassa, crassum
Thick; fleshy, as in *Asarum crassum*

cylindraceus
sil-in-DRA-see-us
cylindracea, cylindraceum

cylindricus
sil-IN-drih-kus
cylindrica, cylindricum
Long and cylindrical, as in *Vaccinium cylindraceum*

deformis
de-FOR-mis
deformis, deforme
Deformed; misshapen, as in *Haemanthus deformis*

demissus
dee-MISS-us
demissa, demissum
Hanging downward; weak, as in *Cytisus demissus*

dendroides
den-DROY-deez

dendroideus
den-DROY-dee-us
dendroidea, dendroideum
Resembling a tree, as in *Sedum dendroideum*

densatus
den-SA-tus
densata, densatum

densus
den-SUS
densa, densum
Compact; dense, as in *Trichodiadema densum*

◖ The twining nature of juvenile English ivy, *Hedera helix*, is referred to in the name *helix*, meaning spiral-shaped. When it reaches maturity as a flowering plant this climbing habit is lost and the ivy becomes a shrub.

implexus
im-PLECK-sus
implexa, implexum
Tangled, as in *Kleinia implexa*

inclinatus
in-klin-AH-tus
inclinata, inclinatum
Bent downward, as in *Moraea inclinata*

indivisus
in-dee-VEE-sus
indivisa, indivisum
Without divisions, as in *Cordyline indivisa*

frutescens
froo-TESS-enz
fruticans
FROO-tih-kanz
fruticosus
froo-tih-KOH-sus
fruticosa, fruticosum
Shrubby; bushy, as in *Argyranthemum frutescens*

intricatus
in-tree-KAH-tus
intricata, intricatum
Tangled, as in *Asparagus intricatus*

difformis
dif-FOR-mis
difformis, difforme
With an unusual form, unlike the rest of the genus, as in *Vinca difformis*

furcans
fur-kanz
furcatus
fur-KA-tus
furcata, furcatum
Forked, as in *Pandanus furcatus*

laxus
LAX-us
laxa, laxum
Loose; open, as in *Freesia laxa*

erectus
ee-RECK-tus
erecta, erectum
Erect; upright, as in *Trillium erectum*

helix
HEE-licks
Spiral-shaped; applied to twining plants, as in *Hedera helix*

lepidus
le-PID-us
lepida, lepidum
Graceful; elegant, as in *Lupinus lepidus*

fastigiatus
fas-tij-ee-AH-tus
fastigiata, fastigiatum
With erect, upright branches, often creating the effect of a column, as in *Cotoneaster fastigiatus*

herbaceus
her-buh-KEE-us
herbacea, herbaceum
Herbaceous (i.e. not woody), as in *Salix herbacea*

lignosus
lig-NOH-sus
lignosa, lignosum
Woody, as in *Tuberaria lignosa*

luxurians
luks-YOO-ee-anz
Luxuriant, as in *Begonia luxurians*

hypogaeus
hy-poh-JEE-us
hypogaea, hypogaeum
Underground; developing in the earth, as in *Copiapoa hypogaea*

natans
NAT-anz
Floating, as in *Trapa natans*

nutans
NUT-anz
Nodding, as in *Billbergia nutans*

obesus
oh-BEE-sus
obesa, obesum
Fat, as in *Euphorbia obesa*

patens
PAT-enz
patulus
PAT-yoo-lus
patula, patulum
With a spreading habit, as in
Salvia patens

pendulinus
pend-yoo-LIN-us
pendulina, pendulinum
Hanging, as in *Salix* × *pendulina*

pendulus
PEND-yoo-lus
pendula, pendulum
Hanging, as in *Betula pendula*

pensilis
PEN-sil-is
pensilis, pensile
Hanging, as in *Glyptostrobus pensilis*

pravissimus
prav-ISS-ih-mus
pravissima, pravissimum
Very crooked, as in *Acacia pravissima*

prolifer
PRO-leef-er
proliferus
pro-LIH-fer-us
prolifera, proliferum
Increasing by the production of sideshoots, as in *Primula prolifera*

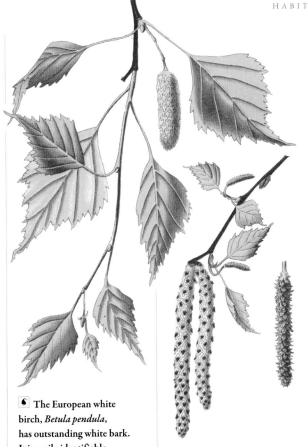

❡ The European white birch, *Betula pendula*, has outstanding white bark. It is easily identifiable by its characteristic pendulous branches.

rectus
REK-tus
recta, rectum
Upright, as in *Phygelius* × *rectus*

robustus
roh-BUS-tus
robusta, robustum
Growing strongly; sturdy, as in *Eremurus robustus*

scandens
SKAN-denz
Climbing, as in *Cobaea scandens*

simplex
SIM-plecks
Simple; without branches, as in *Actaea simplex*

socialis
so-KEE-ah-lis
socialis, sociale
Forming colonies, as in *Crassula socialis*

📍 The tuberous roots of the butterfly weed, *Asclepias tuberosa*, give it a clump-forming habit with numerous upright stems. The roots have also traditionally been used in the treatment of pleurisy, among other things.

solidus
SOL-id-us
solida, solidum
Solid; dense, as in *Corydalis solida*

stans
stanz
Erect; upright, as in *Clematis stans*

strictus
STRIK-tus
stricta, strictum
Erect; upright, as in *Penstemon strictus*

subacaulis
sub-a-KAW-lis
subacaulis, subacaule
Without much stem, as in *Dianthus subacaulis*

subcaulescens
sub-kawl-ESS-enz
With a small stem, as in *Geranium subcaulescens*

suffrutescens
suf-roo-TESS-enz
suffruticosus
suf-roo-tee-KOH-sus
suffruticosa, suffruticosum
Rather shrubby, as in *Paeonia suffruticosa*

suspensus
sus-PEN-sus
suspensa, suspensum
Hanging, as in *Forsythia suspensa*

tenuis
TEN-yoo-is
tenuis, tenue
Slender; thin, as in *Bupleurum tenue*

tenuissimus
ten-yoo-ISS-ih-mus
tenuissima, tenuissimum
Very slender; thin, as in *Stipa tenuissima*

teres
TER-es
With a cylindrical form, as in *Vanda teres*

tortilis
TOR-til-is
tortilis, tortile
Twisted, as in *Acacia tortilis*

tortuosus
tor-tew-OH-sus
tortuosa, tortuosum
Very twisted, as in *Arisaema tortuosum*

tortus
TOR-tus
torta, tortum
Twisted, as in *Masdevallia torta*

tuberosus
too-ber-OH-sus
tuberosa, tuberosum
Tuberous, as in *Polianthes tuberosa*

vescus
VES-kus
vesca, vescum
Thin; feeble, as in *Fragaria vesca*

viminalis
vim-in-AH-lis
viminalis, viminale
vimineus
vim-IN-ee-us
viminea, vimineum
With long, slender shoots, as in *Salix viminalis*

virgatus
vir-GA-tus
virgata, virgatum
Twiggy, as in *Panicum virgatum*

viticella
vy-tee-CHELL-uh
Small vine, as in *Clematis viticella*

volubilis
vol-OO-bil-is
volubilis, volubile
Twining, as in *Aconitum volubile*

In contrast to the herbaceous peonies are the shrubby species known as tree peonies. The habit of moutan, *Paeonia suffruticosa*, is made clear in its name, which means "rather shrubby."

Spreading growth

Plants with a spreading habit can be a blessing and a curse. The neat, weed-suppressing ground cover provided by the rockspray cotoneaster, *Cotoneaster horizontalis*, may be welcome, the relentless growth of creeping buttercup, *Ranunculus repens*, almost certainly will not be. There are a number of names that refer to spreading growth, and coming to understand the nuances in habit which each implies can help us to make wise choices.

diffusus
dy-FEW-sus
diffusa, diffusum
With a spreading habit, as in *Cyperus diffusus*

dilatatus
di-la-TAH-tus
dilatata, dilatatum
Spread out, as in *Dryopteris dilatata*

distortus
DIS-tor-tus
distorta, distortum
Misshapen, as in *Adonis distorta*

divaricatus
dy-vair-ih-KAH-tus
divaricata, divaricatum
With a spreading and straggling habit, as in *Phlox divaricata*

divergens
div-VER-jenz
Spreading out a long way from the center, as in *Ceanothus divergens*

effusus
eff-YOO-sus
effusa, effusum
Spreading loosely, as in *Juncus effusus*

horizontalis
hor-ih-ZON-tah-lis
horizontalis, horizontale
Close to the ground; horizontal, as in *Cotoneaster horizontalis*

humifusus
hew-mih-FEW-sus
humifusa, humifusum
Of a sprawling habit, as in *Opuntia humifusa*

procumbens
pro-KUM-benz
Prostrate, as in *Gaultheria procumbens*

procurrens
pro-KUR-enz
Spreading underground, as in *Geranium procurrens*

prostratus
prost-RAH-tus
prostrata, prostratum
Growing flat on the ground, as in *Veronica prostrata*

repens
REE-penz
With a creeping habit, as in *Gypsophila repens*

reptans
REP-tanz
With a creeping habit, as in *Ajuga reptans*

sarmentosus
sar-men-TOH-sus
sarmentosa, sarmentosum
Producing runners, as in *Androsace sarmentosa*

serpens
SUR-penz
Creeping, as in *Agapetes serpens*

soboliferus
soh-boh-LIH-fer-us
sobolifera, soboliferum
With creeping rooting stems, as in *Geranium soboliferum*

stoloniferus
sto-lon-IH-fer-us
stolonifera, stoloniferum
With runners that take root, as in *Saxifraga stolonifera*

supinus
sup-EE-nus
supina, supinum
Prostrate, as in *Verbena supina*

Dryopteris dilatata

Ajuga reptans

An important distinction

Illustrated are two plants with spreading growth
of different kinds. Bugleweed, *Ajuga reptans*, has
stoloniferous stems that creep over any terrain in sun
or shade. It is so unstoppable that it is often wise to
plant one of its less vigorous variegated cultivars instead.
Meanwhile, the fronds of the wood fern, *Dryopteris
dilatata*, are broad and spreading but are not invasive
in the same way.

Size

Having an idea of the dimensions a plant will ultimately attain offers obvious advantages when planning a garden. A word of caution, however: plant names are often relative and what is considered large or small in one species may not seem so in another. The pretty annual honeywort, *Cerinthe major*, will reach about 2ft in height while the shrub mountain witch alder, *Fothergilla major*, will make about 8ft.

acaulis
el-AH-tus
elata, elatum
Tall, as in *Aralia elata*

altissimus
al-TISS-ih-mus
altissima, altissimum
Very tall; the tallest, as in *Ailanthus altissima*

brevis
BREV-is
brevis, breve
Short, as in *Androsace brevis*

cistena
sis-TEE-nuh
Of dwarf habit, from the Sioux word for baby, as in *Prunus × cistena*

depauperatus
de-por-per-AH-tus
depauperata, depauperatum
Not properly developed; dwarfed, as in *Carex depauperata*

exaltatus
eks-all-TAH-tus
exaltata, exaltatum
Very tall, as in *Nephrolepis exaltata*

excelsior
eks-SEL-see-or
Taller, as in *Fraxinus excelsior*

◀ All things are relative. Masterwort, *Astrantia major*, though larger than the allied small black masterwort, *A. minor*, is a perennial of easily manageable proportions.

excelsus
ek-SEL-sus
excelsa, excelsum
Tall, as in *Metrosideros excelsa*

exiguus
eks-IG-yoo-us
exigua, exiguum
Very little; poor, as in *Salix exigua*

giganteus
jy-GAN-tee-us
gigantea, giganteum
Unusually tall or large, as in *Stipa gigantea*

grossus
GROSS-us
grossa, grossum
Very large, as in *Betula grossa*

humilis
HEW-mil-is
humilis, humile
Low-growing; dwarfish, as in *Chamaerops humilis*

longissimus
lon-JIS-ih-mus
longissima, longissimum
Very long, as in *Aquilegia longissima*

longus
LONG-us
longa, longum
Long, as in *Cyperus longus*

macro-
Used in compound words to denote either long or large

magnus
MAG-nus
magna, magnum
Great; big, as in *Alberta magna*

major
MAY-jor
major, majus
Bigger; larger, as in *Astrantia major*

maximus
MAKS-ih-mus
maxima, maximum
Largest, as in *Rudbeckia maxima*

mega-
Used in compound words to denote big

micro-
Used in compound words to denote small

minimus
MIN-eh-mus
minima, minimum
Smallest, as in *Myosurus minimus*

minor
MY-nor
minor, minus
Smaller, as in *Vinca minor*

minutus
min-YOO-tus
minuta, minutum
Very small, as in *Tagetes minuta*

nanus
NAH-nus
nana, nanum
Dwarf, as in *Betula nana*

⬤ The titan arum, *Amorphophallus titanum*, has the largest inflorescence of any plant and is accompanied by the equally monstrous smell of rotting flesh.

palmetto
pahl-MET-oh
Small palm, as in *Sabal palmetto*

parvus
PAR-vus
parva, parvum
Small, as in *Lilium parvum*

praealtus
pray-AL-tus
praealta, praealtum
Very tall, as in *Symphyotrichum praealtum*

procerus
PRO-ker-us
procera, procerum
Tall, as in *Abies procera*

pumilio
poo-MIL-ee-oh
Small; dwarf, as in *Edraianthus pumilio*

pumilus
POO-mil-us
pumila, pumilum
Dwarf, as in *Trollius pumilus*

pusillus
pus-ILL-us
pusilla, pusillum
Very small, as in *Soldanella pusilla*

pygmaeus
pig-MAY-us
pygmaea, pygmaeum
Dwarf; pygmy, as in *Erigeron pygmaeus*

reductus
red-UK-tus
reducta, reductum
Dwarf, as in *Sorbus reducta*

titanus
ti-AH-nus
titana, titanum
Enormous, as in *Amorphophallus titanum*

Shape

The juxtaposition of shape in the garden is one of the most important ways of making an engaging landscape. The use of diverse forms keeps the eye interested and breaks up the space into a visually stimulating puzzle. But it is not just on the grand scale that shape matters. The shapes comprising each individual plant add to the cohesion and interest of any design.

angularis
ang-yoo-LAH-ris
angularis, angulare
angulatus
ang-yoo-LAH-tus
angulata, angulatum
Angular in shape or form, as in
Jasminum angulare

campanulatus
kam-pan-yoo-LAH-tus
campanulata, campanulatum
In the shape of a bell, as in
Enkianthus campanulatus

cerasiformis
see-ras-if-FOR-mis
cerasiformis, cerasiforme
Shaped like a cherry, as in
Oemleria cerasiformis

circinalis
kir-KIN-ah-lis
circinalis, circinale
Coiled in form, as in *Cycas circinalis*

cochlearis
kok-lee-AH-ris
cochlearis, cochleare
Shaped like a spoon, as in
Saxifraga cochlearis

compressus
kom-PRESS-us
compressa, compressum
Compressed; flattened, as in
Mammillaria compressa

cordatus
kor-DAH-tus
cordata, cordatum
In the shape of a heart, as in
Pontederia cordata

cordifolius
kor-di-FOH-lee-us
cordifolia, cordifolium
With heart-shaped leaves, as in
Crambe cordifolia

cruciatus
kruks-ee-AH-tus
cruciata, crusiatum
In the shape of a cross, as in
Gentiana cruciata

 The drawn-out stems of the ladyfinger cactus, *Mammillaria elongata*, inspired its species epithet, which means lengthened or elongated. The spines of this cactus are harmless so it makes a good houseplant for a bright spot.

cuneatus
kew-nee-AH-tus
cuneata, cuneatum
In the shape of a wedge, as in
Prostanthera cuneata

deltoides
del-TOY-deez
deltoideus
el-TOY-dee-us
deltoidea, deltoideum
Triangular, as in *Dianthus deltoides*

depressus
de-PRESS-us
depressa, depressum
Flattened or pressed down, as in
Gentiana depressa

disciformis
disk-ee-FOR-mis
disciformis, disciforme
Shaped like a disc, as in *Medicago disciformis*

ellipticus
ee-LIP-tih-kus
elliptica, ellipticum
Shaped like an ellipse, as in
Garrya elliptica

elongatus
ee-long-GAH-tus
elongata, elongatum
Lengthened; elongated, as in
Mammillaria elongata

globosus
glo-BOH-sus
globosa, globosum
Round, as in *Buddleja globosa*

globularis
glob-YOO-lah-ris
globularis, globulare
Relating to a small sphere, as in
Carex globularis

gongylodes
GON-jih-loh-deez
Swollen; roundish, as in *Cissus gongylodes*

ligularis
lig-yoo-LAH-ris
ligularis, ligulare
ligulatus
lig-yoo-LAIR-tus
ligulata, ligulatum
Shaped like a strap, as in *Acacia ligulata*

Leaves come in all shapes and sizes, but when we think of a typical leaf it has the silhouette of an ellipse, such as the leaves of the elliptic brush cherry, *Eugenia elliptica*.

obconicus
ob-KON-ih-kus
obconica, obconicum
In the shape of an inverted cone, as in *Primula obconica*

oblatus
ob-LAH-tus
oblata, oblatum
With flattened ends, as in
Syringa oblata

obliquus
oh-BLIK-wus
obliqua, obliquum
Lopsided, as in *Nothofagus obliqua*

🐦 Snowy mespilus,
Amelanchier ovalis, is an
attractive shrub with large,
white flowers and red fruit
which ages to black, but it
is the oval leaves that are
referred to in its
specific name.

oblongatus
ob-long-GAH-tus
oblongata, oblongatum
oblongus
ob-LONG-us
oblonga, oblongum
Oblong, as in *Passiflora
oblongata*

oblongifolius
ob-long-ih-FOH-lee-us
oblongifolia,
oblongifolium
With oblong leaves, as in
Asplenium oblongifolium

obovatus
ob-oh-VAH-tus
obovata, obovatum
In the shape of an inverted egg,
as in *Paeonia obovata*

orbicularis
or-bik-yoo-LAH-ris
orbicularis, orbiculare
orbiculatus
or-bee-kul-AH-tus
orbiculata, orbiculatum
In the shape of a disk; flat and
round, as in *Cotyledon orbiculata*

ovalis
oh-VAH-lis
ovalis, ovale
Oval, as in *Amelanchier ovalis*

ovatus
oh-VAH-tus
ovata, ovatum
Shaped like an egg; ovate, as in
Lagurus ovatus

planus
PLAH-nus
plana, planum
Flat, as in *Eryngium planum*

platy-
Used in compound words to
denote broad (or sometimes flat)

rhombifolius
rom-bih-FOH-lee-us
rhombifolia,
rhombifolium
With diamond-shaped leaves, as
in *Cissus rhombifolia*

sagittalis
saj-ih-TAH-lis
sagittalis, sagittale
sagittatus
saj-ih-TAH-tus
sagittata, sagittatum
Shaped like an arrow, as in
Genista sagittalis

spiralis
spir-AH-lis
spiralis, spirale
Spiral, as in *Macrozamia spiralis*

subcordatus
sub-kor-DAH-tus
subcordata, subcordatum
Shaped rather like a heart, as in
Alnus subcordata

tabularis
tab-yoo-LAH-ris
tabularis, tabulare
tabuliformis
tab-yoo-lee-FORM-is
tabuliformis, tabuliforme
Flat, as in *Blechnum tabulare*

tubiferus
too-BIH-fer-us
tubifera, tubiferum
tubulosus
too-bul-OH-sus
tubulosa, tubulosum
Shaped like a tube or pipe, as in
Clematis tubulosa

Paeonia obovata

Similar but different

The word ovate means egg-shaped; obovate is an inversion of this, with the narrowest part at the base. The names of these two plants both refer to this egg shape. In the case of bunny's tail grass, *Lagurus ovatus*, it is the outline of the inflorescence that is identified, while in the woodland peony, *Paeonia obovata*, it is the leaflets.

Lagurus ovatus

CHAPTER THREE

Features of Plants

When selecting a plant name, botanists often latch onto a certain feature that makes a particular plant stand out from its kin. These names not only tell us something about the species to which they are applied but also act as a handy aide-memoire, focusing on exactly those features that make a plant memorable. Once you know that the species name of mountain fleeceflower, *Persicaria amplexicaulis*, refers to the characteristic clasping of its stems by its leaves, you will always be able to recall it to mind.

Flowers

For many, the uncomplicated pleasure of seeing flowers in abundance is the best reason to grow plants. A walk through a cottage garden in the height of summer opens the door on the astonishing array of floral form that exists. Some of this diversity is reflected in the names that have been given to flowering plants and these help us to catch a glimpse of their horticultural potential.

acmopetala
ak-mo-PET-uh-la
With pointed petals, as in *Fritillaria acmopetala*

acu-
Used in compound words to denote sharply pointed

acuminatus
ah-kew-min-AH-tus
acuminata, acuminatum
Tapering to a long, narrow point, as in *Magnolia acuminata*

adenophorus
ad-eh-NO-for-us
adenophora, adenophorum
With glands, usually in reference to nectar, as in *Salvia adenophora*

apertus
AP-ert-us
aperta, apertum
Open; exposed, as in *Nomocharis aperta*

apetalus
a-PET-uh-lus
apetala, apetalum
Without petals, as in *Sagina apetala*

biflorus
BY-flo-rus
biflora, biflorum
With twin flowers, as in *Polygonatum biflorum*

brachybotrys
brak-ee-BOT-rees
With short clusters, as in *Wisteria brachybotrys*

bracteatus
brak-tee-AH-tus
bracteata, bracteatum

bracteosus
brak-tee-OO-sus
bracteosa, bracteosum

bractescens
brak-TES-senz
With bracts, as in *Veltheimia bracteata*

brevipedunculatus
brev-ee-ped-un-kew-LAH-tus
brevipedunculata, brevipedunculatum
With a short flower stalk, as in *Olearia brevipedunculata*

calcaratus
kal-ka-RAH-tus
calcarata, calcaratum
With spurs, as in *Viola calcarata*

▶ The flowers of the Macartney rose, *Rosa bracteata*, native to China, are surrounded by several large, cut, leaf-like bracts. They give this attractive plant a characteristic appearance.

Corymbs are similar in appearance to the umbels borne by members of the carrot family but, unlike true umbels, here the flowers arise from different points on the stem, as with the blueberry, *Vaccinium corymbosum*.

callianthus
kal-lee-AN-thus
calliantha, callianthum
With beautiful flowers, as in
Berberis calliantha

ciliicalyx
kil-LEE-kal-ux
With a fringed calyx, as in
Menziesia ciliicalyx

confertiflorus
kon-fer-tih-FLOR-us
confertiflora,
confertiflorum
With flowers crowded together,
as in *Salvia confertiflora*

corymbiferus
kor-im-BIH-fer-us
corymbifera,
corymbiferum
With a corymb, as in *Linum corymbiferum*

corymbosus
kor-rim-BOH-sus
corymbosa, corymbosum
With corymbs, as in *Vaccinium corymbosum*

dasystemon
day-see-STEE-mon
With hairy stamens, as in *Tulipa dasystemon*

densiflorus
den-see-FLOR-us
densiflora, densiflorum
Densely flowered, as in
Verbascum densiflorum

discoideus
dis-KOY-dee-us
discoidea, discoideum
Without rays, as in *Matricaria discoidea*

distachyus
dy-STAK-yus
distachya, distachyum
With two spikes, as in *Billbergia distachya*

distylus
DIS-sty-lus
distyla, distylum
With two styles, as in *Acer distylum*

ebracteatus
e-brak-tee-AH-tus
ebracteata, ebracteatum
Without bracts, as in *Eryngium ebracteatum*

emarginatus
e-mar-jin-NAH-tus
emarginata, emarginatum
Slightly notched at the margins,
as in *Pinguicula emarginata*

eriantherus
er-ee-AN-ther-uz
erianthera, eriantherum
With woolly anthers, as in
Penstemon eriantherus

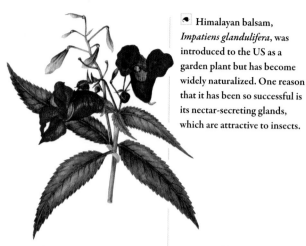

🔹 Himalayan balsam, *Impatiens glandulifera*, was introduced to the US as a garden plant but has become widely naturalized. One reason that it has been so successful is its nectar-secreting glands, which are attractive to insects.

🔹 The common name of balloon flower, *Platycodon grandiflorus*, refers to the swollen shape of the closed buds which open into large, horticulturally valuable flowers, as suggested by its Latin name.

eriostemon
er-ree-oh-STEE-mon
With woolly stamens, as in *Geranium eriostemon*

flabellatus
fla-bel-AH-tus
flabellata, flabellatum
Like an open fan, as in *Aquilegia flabellata*

floribundus
flor-ih-BUN-dus
floribunda, floribunum

floridus
flor-IH-dus
florida, floridum
Very free-flowering, as in *Wisteria floribunda*

flos
Used in combination to denote flower, as in *Lychnis flos-cuculi* (cuckoo flower)

gemmiferus
jem-MIH-fer-us
gemmifera, gemmiferum
With buds, as in *Primula gemmifera*

glanduliferus
glan-doo-LIH-fer-us
glandulifera, glanduliferum
With glands, as in *Impatiens glandulifera*

glomeratus
glom-er-AH-tus
glomerata, glomeratum
With clusters of rounded heads, as in *Campanula glomerata*

graciliflorus
grass-il-ih-FLOR-us
graciliflora, graciliflorum
With slender or graceful flowers, as in *Pseuderanthemum graciliflorum*

grandiflorus
gran-dih-FLOR-us
grandiflora, grandiflorum
With large flowers, as in *Platycodon grandiflorus*

hexandrus
heks-AN-drus
hexandra, hexandrum
With six stamens, as in *Sinopodophyllum hexandrum*

hians
HY-anz
Gaping, as in *Aeschynanthus hians*

hymenanthus
hy-men-AN-thus
hymenantha, hymenanthum
With flowers with a membrane, as in *Trichopilia hymenantha*

involucratus
in-vol-yoo-KRAH-tus
involucrata, involucratum
With a ring of bracts surrounding several flowers, as in *Cyperus involucratus*

jubatus
joo-BAH-tus
jubata, jubatum
With awns, as in *Cortaderia jubata*

📷 The prominent lower lip of the ruby-lipped cattleya, *Cattleya labiata*, is referred to in its scientific name. This feature can be seen in many of the hybrids developed from it for the horticultural trade.

labiatus
la-bee-AH-tus
labiata, labiatum
Lipped, as in *Cattleya labiata*

laetiflorus
lay-tee-FLOR-us
laetiflora, laetiflorum
With bright flowers, as in *Helianthus × laetiflorus*

lasiandrus
las-ee-AN-drus
lasiandra, lasiandrum
With woolly stamens, as in *Clematis lasiandra*

lophanthus
low-FAN-thus
lophantha, lophanthum
With crested flowers, as in *Paraserianthes lophantha*

macranthus
mak-RAN-thus
macrantha, macranthum
With large flowers, as in *Hebe macrantha*

megalanthus
meg-uh-LAN-thus
megalantha, megalanthum
With big flowers, as in *Potentilla megalantha*

monadelphus
mon-ah-DEL-fus
monadelpha, monadelphum
With filaments united, as in *Dianthus monadelphus*

monogynus
mon-NO-gy-nus
monogyna, monogynum
With one pistil, as in *Crataegus monogyna*

monostachyus
mon-oh-STAK-ee-us
monostachya, monostachyum
With one spike, as in *Guzmania monostachya*

nodiflorus
no-dee-FLOR-us
nodiflora, nodiflorum
Flowering at the nodes, as in *Eleutherococcus nodiflorus*

nudiflorus
noo-dee-FLOR-us
nudiflora, nudiflorum
With flowers that appear before the leaves, as in *Jasminum nudiflorum*

octopetalus
ock-toh-PET-uh-lus
octopetala, octopetalum
With eight petals, as in *Dryas octopetala*

pallidiflorus
pal-id-uh-FLOR-us
pallidiflora, pallidiflorum
With pale flowers, as in *Eucomis pallidiflora*

paniculatus
pan-ick-yoo-LAH-tus
paniculata, paniculatum
With flowers arranged in panicles, as in *Koelreuteria paniculata*

peduncularis
pee-dun-kew-LAH-ris
peduncularis, pedunculare
pedunculatus
pee-dun-kew-LA-tus
pedunculata, pedunculatum
With a flower stalk, as in *Lavandula pedunculata*

plenus
plen-US
plena, plenum
Double; full, as in *Sanguinaria canadensis* 'Plena'

polyanthemos
pol-ly-AN-them-os
polyanthus
pol-ee-AN-thus
polyantha, polyanthum
With many flowers, as in *Jasminum polyanthum*

psilostemon
sigh-loh-STEE-mon
With smooth stamens, as in *Geranium psilostemon*

quinqueflorus
kwin-kway-FLOR-rus
quinqueflora, quinqueflorum
With five flowers, as in *Enkianthus quinqueflorus*

racemiflorus
ray-see-mih-FLOR-us
racemiflora, racemiflorum
racemosus
ray-see-MOH-sus
racemosa, racemosum
With flowers that appear in racemes, as in *Nepeta racemosa*

ringens
RIN-jenz
Gaping; open, as in *Arisaema ringens*

sericanthus
ser-ee-KAN-thus
sericantha, sericanthum
With silky flowers, as in *Philadelphus sericanthus*

sexstylosus
seks-sty-LOH-sus
sexstylosa, sexstylosum
With six styles, as in *Hoheria sexstylosa*

spathaceus
spath-ay-SEE-us
spathacea, spathaceum
With a spathe; spathe-like, as in *Salvia spathacea*

sphaerocephalon
sfay-ro-SEF-uh-lon
sphaerocephalus
sfay-ro-SEF-uh-lus
sphaerocephala, sphaerocephalum
With a round head, as in *Allium sphaerocephalon*

spicatus
spi-KAH-tus
spicata, spicatum
With ears that grow in spikes, as in *Mentha spicata*

stylosus
sty-LOH-sus
stylosa, stylosum
With pronounced styles, as in *Rosa stylosa*

tetrandrus
tet-RAN-drus
tetrandra, tetrandrum
With four anthers, as in *Tamarix tetrandra*

triandrus
TRY-an-drus
triandra, triandrum
With three stamens, as in *Narcissus triandrus*

triflorus
TRY-flor-us
triflora, triflorum
With three flowers, as in *Acer triflorum*

umbellatus
um-bell-AH-tus
umbellata, umbellatum
With umbels, as in *Butomus umbellatus*

The highly ornamental Chinese New Year flower, *Enkianthus quinqueflorus*, often has flowers in clusters of five, but there may be as few as three or as many as eight.

47

Leaves

Some plants are immediately recognizable from their leaves. From palm fronds to waterlily pads, foliage is fundamental to a plant's character and essence. In the garden, too, flowers, for all their beauty, are fleeting, and green remains the most important color in the gardener's palette. It is by their leaves that we must learn to know our plants and some names can help us to do this.

actinophyllus
ak-ten-oh-FIL-us
actinophylla,
actinophyllum
With radiating leaves, as in
Schefflera actinophylla

agrifolius
ag-rih-FOH-lee-us
agrifolia, agrifolium
With leaves with a rough or
scabby texture, as in *Quercus
agrifolia*

alternifolius
al-tern-ee-FOH-lee-us
alternifolia, alternifolium
With leaves that grow from
alternating points of a stem,
rather than opposite each other,
as in *Buddleja alternifolia*

angustifolius
an-gus-tee-FOH-lee-us
angustifolia,
angustifolium
With narrow leaves, as in
Pulmonaria angustifolia

aphyllus
a-FIL-us
aphylla, aphyllum
Having, or appearing to have, no
leaves, as in *Asparagus aphyllus*

● Cotoneasters are often
difficult to tell apart but some
species have features that can
provide quick identification.
The name *bullatus* refers to the
characteristic blistered
appearance of the leaves on
hollyberry cotoneaster,
Cotoneaster bullatus.

apiculatus
uh-pik-yoo-LAH-tus
apiculata, apiculatum
With a short, sharp point, as in
Luma apiculata

argutifolius
ar-gew-tih-FOH-lee-us
argutifolia, argutifolium
With sharp-toothed leaves, as in
Helleborus argutifolius

bifolius
by-FOH-lee-us
bifolia, bifolium
With twin leaves, as in *Scilla
bifolia*

bilobatus
by-low-BAH-tus
bilobata, bilobatum
bilobus
by-LOW-bus
biloba, bilobum
With two lobes, as in *Ginkgo
biloba*

bipinnatus
by-pin-NAH-tus
bipinnata, bipinnatum
A leaf that is doubly pinnate, as
in *Cosmos bipinnatus*

biternatus
by-ter-NAH-tus
biternata, biternatum
A leaf that is twice ternate, as in
Actaea biternata

brachyphyllus
brak-ee-FIL-us
brachyphylla,
brachyphyllum
With short leaves, as in
Colchicum brachyphyllum

brevifolius
brev-ee-FOH-lee-us
brevifolia, brevifolium
With short leaves, as in
Gladiolus brevifolius

bullatus
bul-LAH-tus
bullata, bullatum
With blistered or puckered
leaves, as in *Cotoneaster bullatus*

canaliculatus
kan-uh-lik-yoo-LAH-tus
canaliculata,
canaliculatum
With channels or grooves, as in
Erica canaliculata

capillifolius
kap-ill-ih-FOH-lee-us
capillifolia, capillifolium
With hairy leaves, as in
Eupatorium capillifolium

carinatus
kar-IN-uh-tus
carinata, carinatum

cariniferus
kar-in-IH-fer-us
carinifera, cariniferum
With a keel, as in *Allium
carinatum*

centifolius
sen-tih-FOH-lee-us
centifolia, centifolium
With many leaves; with a hundred
leaves, as in *Rosa × centifolia*

The lesser butterfly orchid,
Platanthera bifolia, has an
extremely wide native range,
being found from Ireland in
western Europe to Japan in East
Asia. Its name refers to its
distinctive, twinned leaves.

connatus
kon-NAH-tus
connata, connatum
United; twin; opposite leaves
joined together at the base, as in
Bidens connata

corrugatus
kor-yoo-GAH-tus
corrugata, corrugatum
Corrugated; wrinkled, as in
Salvia corrugata

crassifolius
krass-ih-FOH-lee-us
crassifolia, crassifolium
With thick leaves, as in
Pittosporum crassifolium

crenatus
kre-NAH-tus
crenata, crenatum
Scalloped; crenate, as in *Ilex
crenata*

crispatus
kriss-PAH-tus
crispata, crispatum
crispus
KRISP-us
crispa, crispum
Closely curled, as in *Mentha
crispa*

cuspidatus
kus-pi-DAH-tus
cuspidata, cuspidatum
With a stiff point, as in *Taxus
cuspidata*

decussatus
de-KUSS-ah-tus
decussata, decussatum
With leaves that are borne in
pairs at right angles to each
other, as in *Microbiota decussata*

dentatus
den-TAH-tus
dentata, dentatum
With teeth, as in *Ligularia
dentata*

denticulatus
den-tik-yoo-LAH-tus
denticulata, denticulatum
Slightly toothed, as in *Primula
denticulata*

🔖 The Madeiran orchid, *Dactylorhiza foliosa*, is a native of Madeira named for its lush and healthy foliage. Like many others in its genus, however, it also has impressive racemes of purple flowers.

diphyllus
dy-FIL-us
diphylla, diphyllum
With two leaves, as in *Bulbine diphylla*

dissectus
dy-SEK-tus
dissecta, dissectum
Deeply cut or divided, as in *Cirsium dissectum*

diversifolius
dy-ver-sih-FOH-lee-us
diversifolia, diversifolium
With diverse leaves, as in *Hibiscus diversifolius*

enneaphyllus
en-nee-a-FIL-us
enneaphylla, ennephyllum
With nine leaves or leaflets, as in *Oxalis enneaphylla*

foliolotus
foh-lee-oh-LOH-tus
foliolota, foliolotum
foliolosus
foh-lee-oh-LOH-sus
foliolosa, foliolosum
With leaflets, as in *Thalictrum foliolosum*

foliosus
foh-lee-OH-sus
foliosa, foliosum
With many leaves; leafy, as in *Dactylorhiza foliosa*

grandidentatus
gran-dee-den-TAH-tus
grandidentata, grandidentatum
With big teeth, as in *Thalictrum grandidentatum*

grandifolius
gran-dih-FOH-lee-us
grandifolia, grandifolium
With large leaves, as in *Fagus grandifolia*

heptaphyllus
hep-tah-FIL-us
heptaphylla, heptaphyllum
With seven leaves, as in *Parthenocissus heptaphylla*

hypophyllus
hy-poh-FIL-us
hypophylla, hypophyllum
Underneath the leaf, as in *Ruscus hypophyllum*

impressus
im-PRESS-us
impressa, impressum
With impressed or sunken surfaces, as in *Ceanothus impressus*

incisus
in-KYE-sus
incisa, incisum
With deeply cut and irregular incisions, as in *Prunus incisa*

integrifolius
in-teg-ree-FOH-lee-us
integrifolia, integrifolium
With leaves that are complete; uncut, as in *Meconopsis integrifolia*

isophyllus
eye-so-FIL-us
isophylla, isophyllum
With leaves of the same size, as in *Penstemon isophyllus*

laciniatus
la-sin-ee-AH-tus
laciniata, laciniatum
Divided into narrow sections, as in *Rudbeckia laciniata*

latifrons
lat-ee-FRONS
With broad fronds, as in *Encephalartos latifrons*

leiophyllus
lay-oh-FIL-us
leiophylla, leiophyllum
With smooth leaves, as in *Pinus leiophylla*

linearis
lin-AH-ris
linearis, lineare
With narrow, almost parallel sides, as in *Ceropegia linearis*

macrodontus
mak-roh-DON-tus
macrodonta, macrodontum
With large teeth, as in *Olearia macrodonta*

maculatus
mak-yuh-LAH-tus
maculata, maculatum

maculosus
mak-yuh-LAH-sus
maculosa, maculosum
With spots, as in *Begonia maculata*

millefoliatus
mil-le-foh-lee-AH-tus
millefoliata, millefoliatum

millefolius
mil-le-FOH-lee-us
millefolia, millefolium
With many leaves (literally a thousand leaves), as in *Achillea millefolium*

multifidus
mul-TIF-id-us
multifida, multifidum
With many divisions, usually of leaves with many tears, as in *Helleborus multifidus*

multisectus
mul-tee-SEK-tus
multisecta, multisectum
With many cuts, as in *Geranium multisectum*

myriophyllus
mir-ee-oh-FIL-us
myriophylla, myriophyllum
With very many leaves, as in *Acaena myriophylla*

nervis
NERV-is
nervis, nerve

nervosus
ner-VOH-sus
nervosa, nervosum
With visible nerves, as in *Astelia nervosa*

non-scriptus
non- SKRIP-tus
non-scripta, non-scriptum
Without any markings, as in *Hyacinthoides non-scripta*

obtusifolius
ob-too-sih-FOH-lee-us
obtusifolia, obtusifolium
With blunt leaves, as in *Peperomia obtusifolia*

oppositifolius
op-po-sih-tih-FOH-lee-us
oppositifolia, oppositifolium
With leaves that grow opposite each other from the stem, as in *Chiastophyllum oppositifolium*

palmatus
pahl-MAH-tus
palmata, palmatum
Palmate, as in *Acer palmatum*

perfoliatus
per-foh-lee-AH-tus
perfoliata, perfoliatum
With the leaf surrounding the stem, as in *Parahebe perfoliata*

perforatus
per-for-AH-tus
perforata, perforatum
With, or appearing to have, small holes, as in *Hypericum perforatum*

◀ Although it has the appearance of an exotic succulent, lamb's tail, *Chiastophyllum oppositifolium*, is actually hardy. Its name refers to its leaves, which are in pairs along the stems.

perulatus
per-uh-LAH-tus
perulata, perulatum
With perules (bud scales), as in
Enkianthus perulatus

petiolaris
pet-ee-OH-lah-ris
petiolaris, petiolare

petiolatus
pet-ee-oh-LAH-tus
petiolata, petiolatum
With a leaf stalk, as in
Helichrysum petiolare

picturatus
pik-tur-AH-tus
picturata, picturatum
With variegated leaves, as in
Calathea picturata

pinguifolius
pin-gwih-FOH-lee-us
pinguifolia, pinguifolium
With fat leaves, as in *Hebe*
pinguifolia

planifolius
plan-ih-FOH-lee-us
planifolia, planifolium
With flat leaves, as in *Iris planifolia*

plicatus
ply-KAH-tus
plicata, plicatum
Pleated, as in *Thuja plicata*

podophyllus
po-do-FIL-us
podophylla, podophyllum
With stout-stalked leaves, as in
Rodgersia podophylla

porophyllus
po-ro-FIL-us
porophylla, porophyllum
With leaves with (apparent)
holes, as in *Saxifraga porophylla*

quinquefolius
kwin-kway-FOH-lee-us
quinquefolia,
quinquefolium
With five leaves, often referring
to leaflets, as in *Parthenocissus*
quinquefolia

revolutus
re-vo-LOO-tus
revoluta, revolutum
Rolled backwards (e.g. of leaves),
as in *Cycas revoluta*

rhytidophyllus
ry-ti-do-FIL-us
rhytidophylla,
rhytidophyllum
With wrinkled leaves, as in
Viburnum rhytidophyllum

**Most *Iris* species have fans
of tightly folded leaves but
those of the flat-leaved iris,
I. planifolia, are different, being
opened out and spreading.**

rotundifolius
ro-tun-dih-FOH-lee-us
rotundifolia,
rotundifolium
With leaves that are round, as in
Prostanthera rotundifolia

rugosus
roo-GOH-sus
rugosa, rugosum
Wrinkled, as in *Rosa rugosa*

serratifolius
sair-rat-ih-FOH-lee-us
serratifolia, serratifolium
With leaves that are serrated or
saw-toothed, as in *Photinia*
serratifolia

serratus
sair-AH-tus
serrata, serratum
With small-toothed leaf
margins, as in *Zelkova serrata*

serrulatus
ser-yoo-LAH-tus
serrulata, serrulatum,
With small serrations at the leaf
margins, as in *Enkianthus*
serrulatus

simplicifolius
sim-plik-ih-FOH-lee-us
simplicifolia,
simplicifolium
With simple leaves, as in *Astilbe*
simplicifolia

squamaria
SKWA-ma-ria
squamarius, squamarium
Scale-clad, covered with scales,
as in *Lathraea squamaria*

squamatus
SKWA-ma-tus
squamata, squamatum
With small, scale-like leaves or
bracts, as in *Juniperus squamata*

stipulaceus
stip-yoo-LAY-see-us
stipulacea, stipulaceum

stipularis
stip-yoo-LAH-ris
stipularis, stipulare

stipulatus
stip-yoo-LAH-tus
stipulata, stipulatum
With stipules, as in *Oxalis
stipularis*

tenuifolius
ten-yoo-ih-FOH-lee-us
tenuifolia, tenuifolium
With slender leaves, as in
Pittosporum tenuifolium

ternatus
ter-NAH-tus
ternata, ternatum
With clusters of three, as in
Choisya ternata

tricuspidatus
try-kusp-ee-DAH-tus
tricuspidata,
tricuspidatum
With three points, as in
Parthenocissus tricuspidata

trifoliatus
try-foh-lee-AH-tus
trifoliata, trifoliatum

trifolius
try-FOH-lee-us
trifolia, trifolium
With three leaves, as in *Gillenia
trifoliata*

◖ The leaves of the
Boston ivy, *Parthenocissus
tricuspidata*, are
distinctively three-pointed
and this is perfectly
reflected in its scientific
name. Its common name,
however, is something of a
misnomer as it is native to
East Asia.

trigonophyllus
try-gon-oh-FIL-us
trigonophylla,
trigonophyllum
With triangular leaves, as in
Acacia trigonophylla

trinervis
try-NER-vis
trinervis, trinerve
With three nerves, as in
Coelogyne trinervis

triplinervis
trip-lin-NER-vis
triplinervis, triplinerve
With three veins, as in *Anaphalis
triplinervis*

triternatus
try-tern-AH-tus
triternata, triternatum
Literally three threes, referring to
leaf shape, as in *Corydalis
triternata*

undatus
un-DAH-tus
undata, undatum

undulatus
un-dew-LAH-tus
undulata, undulatum
Wavy; undulating, as in *Hosta
undulata*

variegatus
var-ee-GAH-tus
variegata, variegatum
Variegated, as in *Pleioblastus
variegatus*

venosus
ven-OH-sus
venosa, venosum
With many veins, as in *Vicia
venosa*

zonalis
zo-NAH-lis
zonalis, zonale

zonatus
zo-NAH-tus
zonata, zonatum
With bands, often colored, as in
Cryptanthus zonatus

Fruit

As the gardening year draws to its close and the first frosts snap at the final flowers, fruit can be an unanticipated extra. After the blowsy splendor of rose, *Rosa*, flowers has faded may come the more subtle pleasure of rose hips. When the spring flowers and fall foliage of rowan, *Sorbus aucuparia*, are just a memory, their fruit can still lift the spirits. Some plant names can help draw attention to unsuspected fall delights.

aggregatus
ag-gre-GAH-tus
aggregata, aggregatum
Denotes aggregate flowers or fruits, such as raspberry or strawberry, as in *Eucalyptus aggregata*

baccans
BAK-kanz
bacciferus
bak-IH-fer-us
baccifera, bacciferum
With berries, as in *Erica baccans*

callicarpus
kal-ee-KAR-pus
callicarpa, callicarpum
With beautiful fruit, as in *Sambucus callicarpa*

cocciferus
koh-KIH-fer-us
coccifera, cocciferum
coccigerus
koh-KEE-ger-us
coccigera, coccigerum
Producing berries, as in *Eucalyptus coccifera*

coniferus
koh-NIH-fer-us
conifera, coniferum
With cones, as in *Magnolia conifera*

drupaceus
droo-PAY-see-us
drupacea, drupaceum
drupiferus
droo-PIH-fer-us
drupifera, drupiferum
With fleshy fruits, such as peach or cherry, as in *Hakea drupacea*

The pekea nut, *Caryocar nuciferum*, is a South American tree. The nut-like fruit, which are said to taste of almonds, are about the size of a coconut, *Cocos nucifera*, with which the tree shares its species name, meaning "producing nuts."

holocarpus
ho-loh-KAR-pus
holocarpa, holocarpum
With complete fruit, as in
Staphylea holocarpa

ixocarpus
iks-so-KAR-pus
ixocarpa, ixocarpum
With sticky fruit, as in *Physalis ixocarpa*

lappa
LAP-ah
A bur (prickly seed case or flowerhead), as in *Arctium lappa*

macrospermus
mak-roh-SPERM-us
macrosperma, macrospermum
With large seeds, as in *Senecio macrospermus*

monopyrenus
mon-NO-py-ree-nus
monopyrena, monopyrenum
With a single stone or pit, as in *Cotoneaster monopyrenus*

myriocarpus
mir-ee-oh-KAR-pus
myriocarpa, myriocarpum
With very many fruits, as in *Schefflera myriocarpa*

nucifer
NOO-siff-er
nucifera, nuciferum
Producing nuts, as in *Cocos nucifera*

pileatus
py-lee-AH-tus
pileata, pileatum
With a cap, as in *Lonicera pileata*

pilularis
pil-yoo-LAH-ris
pilularis, pilulare
piluliferus
pil-loo-LIH-fer-us
pilulifera, piluliferum
With globular fruit, as in *Urtica pilulifera*

prolificus
pro-LIF-ih-kus
prolifica, prolificum
Producing many fruits, as in *Echeveria prolifica*

❝ The woody pods of kowhai, *Sophora tetraptera*, are characteristically crested with four wings, signified by its name.

tetrapterus
tet-rap-TER-us
tetraptera, tetrapterum
With four wings, as in *Sophora tetraptera*

trachyspermus
trak-ee-SPER-mus
trachysperma, trachyspermum
With rough seeds, as in *Sauropus trachyspermus*

Stems and shoots

A surprising number of plant names refer to stems and shoots. These may seem irrelevant characteristics but they should not be discounted. Most of the ornamental traits associated with stems appear in other sections (*albosinensis*, for example, which refers to the bleached trunks of the white Chinese birch, *Betula albosinensis*, is in Light colors) but, nevertheless, the structural elements of plants contribute much to their overall appearance.

alatus
a-LAH-tus
alata, alatum
Winged, as in *Euonymus alatus*

amplexicaulis
am-pleks-ih-KAW-lis
amplexicaulis,
amplexicaule
Clasping the stem, as in
Persicaria amplexicaulis

articulatus
ar-tik-oo-LAH-tus
articulata, articulatum
With a jointed stem, as in
Senecio articulatus

bulbiferus
bulb-IH-fer-us
bulbifera, bulbiferum
bulbiliferus
bulb-il-IH-fer-us
bulbilifera, bulbiliferum
With bulbs, often referring to
bulbils, as in *Lachenalia*
bulbifera

bulbocodium
bulb-oh-KOD-ee-um
With a woolly bulb, as in
Narcissus bulbocodium

capillipes
cap-ILL-ih-peez
With slender feet, as in *Acer*
capillipes

crassicaulis
krass-ih-KAW-lis
crassicaulis, crassicaule
With a thick stem, as in *Begonia*
crassicaulis

crassipes
KRASS-ih-peez
With thick feet or thick stems,
as in *Quercus crassipes*

decurrens
de-KUR-enz
Running down the stem, as in
Calocedrus decurrens

filipes
fil-EE-pays
With thread-like stalks, as in
Rosa filipes

fistulosus
fist-yoo-LOH-sus
fistulosa, fistulosum
Hollow, as in *Asphodelus*
fistulosus

flexicaulis
fleks-ih-KAW-lis
flexicaulis, flexicaule
With a supple stem, as in
Strobilanthes flexicaulis

flexilis
FLEKS-il-is
flexilis, flexile
Pliant, as in *Pinus flexilis*

flexuosus
fleks-yoo-OH-sus
flexuosa, flexuosum
Indirect; zigzagging, as in
Corydalis flexuosa

fragilis
FRAJ-ih-lis
fragilis, fragile
Brittle; quick to wilt, as in *Salix*
fragilis

gracilipes
gra-SIL-i-peez
With a slender stalk, as in
Mahonia gracilipes

hirtipes
her-TYE-pees
With hairy stems, as in *Viola*
hirtipes

longicaulis
lon-jee-KAW-lis
longicaulis, longicaule
With long stalks, as in
Aeschynanthus longicaulis

longipes
LON-juh-peez
With a long stalk, as in *Acer longipes*

medullaris
med-yoo-LAH-ris
medullaris, medullare
medullus
med-DUL-us
medulla, medullum
Pithy, as in *Cyathea medullaris*

nodosus
nod-OH-sus
nodosa, nodosum
With conspicuous joints or nodes, as in *Geranium nodosum*

nudicaulis
new-dee-KAW-lis
nudicaulis, nudicaule
With bare stems, as in *Papaver nudicaule*

planipes
PLAN-ee-pays
With a flat stalk, as in *Euonymus planipes*

quadrangularis
kwad-ran-gew-LAH-ris
quadrangularis, quadrangulare
quadrangulatus
kwad-ran-gew-LAH-tus
quadrangulata, quadrangulatum
With four angles, as in *Passiflora quadrangularis*

radicans
RAD-ee-kanz
With stems that take root, as in *Campsis radicans*

ramosus
ram-OH-sus
ramosa, ramosum
Branched, as in *Anthericum ramosum*

rigens
RIG-enz
rigidus
RIG-ih-dus
rigida, rigidum
Rigid; inflexible; stiff, as in *Verbena rigida*

sarcocaulis
sar-koh-KAW-lis
sarcocaulis, sarcocaule
With a fleshy stem, as in *Crassula sarcocaulis*

scaposus
ska-POH-sus
scaposa, scaposum
With leafless flowering stems (scapes) as in *Aconitum scaposum*

sessili-
Used in compound words to denote stalkless

◄ Known for its fiery fall foliage, the burning bush, *Euonymus alatus*, is also distinctive for the corky wings that ridge its stems.

tenuicaulis
ten-yoo-ee-KAW-lis
tenuicaulis, tenuicaule
With slender stems, as in *Dahlia tenuicaulis*

trichotomus
try-KOH-toh-mus
trichotoma, trichotomum
With three branches, as in *Clerodendrum trichotomum*

ventricosus
ven-tree-KOH-sus
ventricosa, ventricosum
With a swelling on one side, belly-like, as in *Ensete ventricosum*

Texture and thorns

One of the most enjoyable ways of experiencing plants is to move among them, feeling their leaves and brushing aside their branches. This quality in plants is acknowledged in the modern idea of the Sensory Garden. Not every plant is suitable for this treatment, however, especially those that bear thorns. Fortunately, there are a number of names that forewarn of these potential hazards.

aculeatus
a-kew-lee-AH-tus
aculeata, aculeatum
Prickly, as in *Polystichum aculeatum*

armatus
arm-AH-tus
armata, armatum
With thorns, spines, or spikes, as in *Dryandra armata*

asper
AS-per
aspera, asperum

asperatus
as-per-AH-tus
asperata, asperatum
With a rough texture, as in *Hydrangea aspera*

asperrimus
as-PER-rih-mus
asperrima, asperrimum
With a very rough texture, as in *Agave asperrima*

barbinervis
bar-bih-NER-vis
barbinervis, barbinerve
With bearded or barbed veins, as in *Clethra barbinervis*

callosus
kal-OH-sus
callosa, callosum
With thick skin; with calluses, as in *Saxifraga callosa*

▪ The specific epithet of prickly dryandra, *Dryandra armata*, offers a warning to gardeners that it should be handled with care.

calvus
KAL-vus
calva, calvum
Without hair; naked, as in *Viburnum calvum*

coriaceus
kor-ee-uh-KEE-us
coriacea, coriaceum
Thick, tough, and leathery, as in *Paeonia coriacea*

crinitus
krin-EE-tus
crinita, crinitum
With long, weak hairs, as in *Acanthophoenix crinita*

farinaceus
far-ih-NAH-kee-us
farinacea, farinaceum
Producing starch; mealy, like flour, as in *Salvia farinacea*

fibrillosus
fy-BRIL-oh-sus
fibrillosa, fibrillosum
fibrosus
fy-BROH-sus
fibrosa, fibrosum
Fibrous, as in *Dicksonia fibrosa*

floccigerus
flok-KEE-jer-us
floccigera, floccigerum
floccosus
flok-KOH-sus
floccosa, floccosum
With a woolly texture, as in *Rhipsalis floccosa*

glabellus
gla-BELL-us
glabella, glabellum
Smooth, as in *Epilobium glabellum*

glaber
GLAY-ber
glabra, glabrum
Smooth, hairless, as in *Bougainvillea glabra*

glabratus
GLAB-rah-tus
glabrata, glabratum

glabrescens
gla-BRES-senz

glabriusculus
gla-bree-US-kyoo-lus
glabriuscula, glabriusculum
Rather hairless, as in *Corylopsis glabrescens*

heteracanthus
het-er-a-KAN-thus
heteracantha, heteracanthum
With various or diverse spines, as in *Agave heteracantha*

hirsutus
her-SOO-tus
hirsuta, hirsutum
Hairy, as in *Lotus hirsutus*

hirtellus
her-TELL-us
hirtella, hirtellum
Rather hairy, as in *Plectranthus hirtellus*

hirtus
HER-tus
hirta, hirtum
Hairy, as in *Columnea hirta*

hispidus
HISS-pih-dus
hispida, hispidum
With bristles, as in *Leontodon hispidus*

horridus
HOR-id-us
horrida, horridum
With many thorns, as in *Euphorbia horrida*

inermis
IN-er-mis
inermis, inerme
Without arms (e.g. without thorns), as in *Acaena inermis*

laevigatus
lee-vih-GAH-tus
laevigata, laevigatum

laevis
LEE-vis
laevis, laeve
Smooth, as in *Crocus laevigatus*

lanatus
la-NA-tus
lanata, lanatum
Woolly, as in *Lavandula lanata*

❛ The shining, hairless leaves and fruit of the Barbados cherry, *Malpighia glabra*, are referred to in its name. It is a tropical tree primarily grown as an ornamental, but which bears edible fruit that have very high levels of vitamin C.

lanigerus
lan-EE-ger-rus
lanigera, lanigerum

lanosus
LAN-oh-sus
lanosa, lanosum

lanuginosus
lan-oo-gih-NOH-sus
lanuginosa, lanuginosum
Woolly, as in *Leptospermum lanigerum*

lasioglossus
las-ee-oh-GLOSS-us
lasioglossa, lasioglossum
With a rough tongue, as in *Lycaste lasioglossa*

malacoides
mal-a-KOY-deez
Soft, as in *Erodium malacoides*

One of the associations that springs most readily to mind when we think of roses are their thorns, but the epithet of the Scotch rose, *Rosa spinosissima*, warns of a particularly thick covering of tiny thorns.

micranthus
mi-KRAN-thus
micrantha, micranthum
With small flowers, as in
Heuchera micrantha

mollis
MAW-lis
mollis, molle
Soft; with soft hairs, as in
Alchemilla mollis

mollissimus
maw-LISS-ih-mus
mollissima, mollissimum
Very soft, as in *Passiflora mollissima*

muricatus
mur-ee-KAH-tus
muricata, muricatum
With rough and hard points, as in *Solanum muricatum*

pilosus
pil-OH-sus
pilosa, pilosum
With long, soft hairs, as in
Symphyotrichum pilosum

pocophorus
po-KO-for-us
pocophora, pocophorum
Fleece-bearing, as in
Rhododendron pocophorum

polytrichus
pol-ee-TRY-kus
polytricha, polytrichum
With many hairs, as in *Thymus polytrichus*

pubens
PEW-benz

pubescens
pew-BESS-enz
Downy, as in *Primula ×
pubescens*

pubigerus
pub-EE-ger-us
pubigera, pubigerum
Producing soft hairs, as in
Schefflera pubigera

pycnacanthus
pik-na-KAN-thus
pycnacantha,
pycnacanthum
Densely spined, as in
Coryphantha pycnacantha

scaber
SKAB-er
scabra, scabrum
Rough, as in *Eccremocarpus scaber*

sericeus
ser-IK-ee-us
sericea, sericeum
Silky, as *Rosa sericea*

setaceus
se-TAY-see-us
setacea, setaceum
With bristles, as in *Pennisetum setaceum*

setiferus
set-IH-fer-us
setifera, setiferum
With bristles, as in *Polystichum setiferum*

spinescens
spy-NES-enz
spinifex
SPIN-ee-feks
spinosus
spy-NOH-sus
spinosa, spinosum
With spines, as in *Acanthus spinosus*

spinosissimus
spin-oh-SIS-ih-mus
spinosissima, spinosissimum
Very spiny, as in *Rosa spinosissima*

strigosus
strig-OH-sus
strigosa, strigosum
With stiff bristles, as in *Rubus strigosus*

subhirtellus
sub-hir-TELL-us
subhirtella, subhirtellum
Rather hairy, as in *Prunus × subhirtella*

subtomentosus
sub-toh-men-TOH-sus
subtomentosa, subtomentosum
Almost hairy, as in *Rudbeckia subtomentosa*

tenax
TEN-aks
Tough; matted, as in *Phormium tenax*

tomentosus
toh-men-TOH-sus
tomentosa, tomentosum
Very woolly; matted, as in *Paulownia tomentosa*

triacanthos
try-a-KAN-thos
With three spines, as in *Gleditsia triacanthos*

vernicifluus
ver-nik-IF-loo-us
verniciflua, vernicifluum
Producing varnish, as in *Rhus verniciflua*

vernicosus
vern-ih-KOH-sus
vernicosa, vernicosum
Varnished, as in *Hebe vernicosa*

villosus
vil-OH-sus
villosa, villosum
With soft hairs, as in *Photinia villosa*

Eccremocarpus scaber

Lycaste lasioglossa

BEHIND THE NAME
Roughly speaking

The purpose of botanical names is to assist us in understanding plants, yet sometimes they can leave us curious. The epithet *scaber* means rough but it would require a good guess to identify that in the Chilean glory flower, *Eccremocarpus scaber,* this refers to the seed pods. Other names are more specific: the epithet *lasioglossa*, as in the shaggy-lipped lycaste, *Lycaste lasioglossa*, refers to the roughness of the tongue-like lip of the flower.

Taste and smell

A pleasure that all true gardeners know is that of flipping through a book of plants and growing in imagination those pictured. But there are some features of plants that an image will not tell, namely their fragrance and flavor. Knowing that the perfume of winter daphne, *Daphne odora*, is sweet and lovely or that the foliage of Chilean box thorn, *Vestia foetida*, is unpleasantly pungent may influence your decision whether to grow them.

acer
AY-sa
acris, acre
With a sharp or pungent taste, as in *Sedum acre*

acetosella
a-kee-TOE-sell-uh
With slightly sour leaves, as in *Oxalis acetosella*

aethusifolius
e-thu-si-FOH-lee-us
aethusifolia, aethusifolium
With pungent leaves like those of *Aethusa*, as in *Aruncus aethusifolius*

ananassa
a-NAN-ass-uh
ananassae
a-NAN-ass-uh-ee
With a fragrance like pineapple, as in *Fragaria* × *ananassa*

aromaticus
ar-oh-MAT-ih-kus
aromatica, aromaticum
With a fragrant, aromatic scent, as in *Lycaste aromatica*

assa-foetida
ASS-uh FET-uh-duh
From the Persian *aza*, mastic, and Latin *foetidus*, stinking, as in *Ferula assa-foetida*

balsameus
bal-SAM-ee-us
balsamea, balsameum
Like balsam, as in *Abies balsamea*

◀ The almond, *Prunus dulcis*, is well known for its sweet-tasting seeds, a property referred to in its name. Bitter almonds belong to the same species but are recognized as var. *amara*, which means bitter.

camphora
kam-for-AH
camphoratus
kam-for-AH-tus
camphorata, camphoratum
Like camphor, as in *Thymus camphoratus*

citriodorus
sit-ree-oh-DOR-us
citriodora, citriodorum
With the scent of lemons, as in *Thymus citriodorus*

deliciosus
de-lis-ee-OH-sus
deliciosa, deliciosum
Delicious, as in *Monstera deliciosa*

dulcis
DUL-sis
dulcis, dulce
Sweet, as in *Prunus dulcis*

edulis
ED-yew-lis
edulis, edule
Edible, as in *Dioon edule*

esculentus
es-kew-LEN-tus
esculenta, esculentum
Edible, as in *Colocasia esculenta*

foetidissimus
fet-uh-DISS-ih-mus
foetidissima, foetidissimum
With a very bad smell, as in *Iris foetidissima*

foetidus

FET-uh-dus
foetida, foetidum
With a bad smell, as in *Vestia foetida*

fragrans

FRAY-granz
Fragrant, as in *Osmanthus fragrans*

fragrantissimus

fray-gran-TISS-ih-mus
fragrantissima, fragrantissimum
Very fragrant, as in *Lonicera fragrantissima*

graveolens

grav-ee-OH-lenz
With a heavy scent, as in *Ruta graveolens*

inodorus

in-oh-DOR-us
inodora, inodorum
Without scent, as in *Hypericum × inodorum*

melliferus

mel-IH-fer-us
mellifera, melliferum
Producing honey, as in *Euphorbia mellifera*

melliodorus

mel-EE-uh-do-rus
melliodora, melliodorum
With the scent of honey, as in *Eucalyptus melliodora*

moschatus

MOSS-kuh-tus
moschata, moschatum
Musky, as in *Malva moschata*

Often grown in herb gardens for its aromatic leaves, rue, *Ruta graveolens*, has a number of traditional uses, such as in the treatment of hysteria. Principally an ornamental plant now, care must be taken with it as its foliage can cause blistering.

odoratissimus

oh-dor-uh-TISS-ih-mus
odoratissima, odoratissimum
With a very fragrant scent, as in *Viburnum odoratissimum*

odoratus

oh-dor-AH-tus
odorata, odoratum

odoriferus

oh-dor-IH-fer-us
odorifera, odoriferum

odorus

oh-DOR-us
odora, odorum
With a fragrant scent, as in *Lathyrus odoratus*

piperitus

pip-er-EE-tus
piperita, piperitum
With a pepper-like taste, as in *Mentha × piperita*

sapidus

sap-EE-dus
sapida, sapidum
With a pleasant taste, as in *Rhopalostylis sapida*

suaveolens

swah-vee-OH-lenz
With a sweet fragrance, as in *Brugmansia suaveolens*

unedo

YOO-nee-doe
Edible but of doubtful taste, from *unum edo*, I eat one, as in *Arbutus unedo*

Additional features

Plants are so multifarious in their characteristics that some of the names applied to them are difficult to categorize. In addition, some names are ambiguous with regards to the part of the plant to which they refer. Nevertheless, knowing that *marmoratum* means marbled or that *nitens* means shining perhaps gives some impression of a plant's appearance. Other less evocative names may simply stimulate our curiosity.

adpressus
ad-PRESS-us
adpressa, adpressum
Pressed close to; refers to the way hairs (for example) press against a stem, as in *Cotoneaster adpressus*

appressus
a-PRESS-us
appressa, appressum
Pressed close against, as in *Carex appressa*

apterus
AP-ter-us
aptera, apterum
Without wings, as in *Odontoglossum apterum*

aristatus
a-ris-TAH-tus
aristata, aristatum
Bearded, as in *Aloe aristata*

attenuatus
at-ten-yoo-AH-tus
attenuata, attenuatum
With a narrow point, as in *Haworthia attenuata*

axillaris
ax-ILL-ah-ris
axillaris, axillare
Growing in the axil, as in *Petunia axillaris*

◧ The wax plant, *Hoya carnosa*, is a climbing plant from warm parts of Asia grown in houses for its fragrant flowers. Its name refers to the thick, hairless, fleshy texture of its leaves and flowers.

bicornis
BY-korn-is
bicornis, bicorne
bicornutus
by-kor-NOO-tus
bicornuta, bicornutum
With two horns or horn-like spurs, as in *Passiflora bicornis*

bifidus
BIF-id-us
bifida, bifidum
Cleft in two parts, as in *Rhodophiala bifida*

brachycerus
brak-ee-SER-us
brachycera, brachycerum
With short horns, as in *Gaylussacia brachycera*

capitatus
kap-ih-TAH-tus
capitata, capitatum
Flowers, fruit, or whole plant growing in a dense head, as in *Cornus capitata*

capreolatus
kap-ree-oh-LAH-tus
capreolata, capreolatum
With tendrils, as in *Bignonia capreolata*

carnosus
kar-NOH-sus
carnosa, carnosum
Fleshy, as in *Hoya carnosa*

cavus
KA-vus
cava, cavum
Hollow, as in *Corydalis cava*

ciliaris
sil-ee-AH-ris
ciliaris, ciliare

ciliatus
sil-ee-ATE-us
ciliata, ciliatum
With leaves and petals that are
fringed with hairs, as in
Tropaeolum ciliatum

cirratus
sir-RAH-tus
cirrata, cirratum

cirrhosus
sir-ROH-sus
cirrhosa, cirrhosum
With tendrils, as in *Clematis
cirrhosa*

conjunctus
kon-JUNK-tus
conjuncta, conjunctum
Joined, as in *Alchemilla conjuncta*

constrictus
kon-STRIK-tus
constricta, constrictum
Constricted, as in *Yucca
constricta*

copallinus
kop-al-EE-nus
copallina, copallinum
With gum or resin, as in *Rhus
copallinum*

costatus
kos-TAH-tus
costata, costatum
With ribs, as in *Aglaonema
costatum*

The silver wattle, *Acacia
dealbata*, **is a fast-growing tree.
Attractive in flower and foliage,
its name makes reference to the
white dusting on its leaves.**

cristatus
kris-TAH-tus
cristata, cristatum
With tassel-like tips, as in *Iris
cristata*

crustatus
krus-TAH-tus
crustata, crustatum
Encrusted, as in *Saxifraga
crustata*

dealbatus
day-al-BAH-tus
dealbata, dealbatum
Covered with an opaque white
powder, as in *Acacia dealbata*

deflexus
de-FLEKS-us
deflexa, deflexum
Bending sharply downward, as in
Enkianthus deflexus

denudatus
dee-noo-DAH-tus
denudata, denudatum
Bare; naked, as in *Magnolia
denudata*

didymus
DID-ih-mus
didyma, didymum
In pairs; twin, as in *Monarda
didyma*

distichus
DIS-tih-kus
disticha, distichum
In two ranks or levels, as in
Taxodium distichum

excorticatus
eks-kor-tih-KAH-tus
excorticata, excorticatum
Lacking or stripped of bark, as in
Fuchsia excorticata

fascicularis
fas-sik-yoo-LAH-ris
fascicularis, fasciculare

fasciculatus
fas-sik-yoo-LAH-tus
fasciculata, fasciculatum
Clustered or grouped together in
bundles, as in *Ribes fasciculatum*

filamentosus
fil-uh-men-TOH-sus
filamentosa, filamentosum

filarius
fil-AH-ree-us
filaria, filarium
With filaments or threads, as in
Yucca filamentosa

fimbriatus
fim-bry-AH-tus
fimbriata, fimbriatum
Fringed, as in *Silene fimbriata*

fissilis
FISS-ill-is
fissilis, fissile

fissus
FISS-us
fissa, fissum

fissuratus
fis-zhur-RAH-tus
fissurata, fissuratum
With a split, as in *Alchemilla fissa*

flaccidus
FLA-sih-dus
flaccida, flaccidum
Weak: soft; feeble, as in *Yucca flaccida*

foveolatus
foh-vee-oh-LAH-tus
foveolata, foveolatum
With slight pitting, as in
Chionanthus foveolatus

fugax
FOO-gaks
Withering quickly; fleeting, as in
Urginea fugax

fulgens
FUL-jenz

fulgidus
FUL-jih-dus
fulgida, fulgidum
Shining; glistening, as in
Rudbeckia fulgida

gibbosus
gib-OH-sus
gibbosa, gibbosum

gibbus
GIB-us
gibba, gibbum
With a swelling on one side, as
in *Fritillaria gibbosa*

globiferus
glo-BIH-fer-us
globifera, globiferum
With spherical clusters of small
globes, as in *Pilularia globifera*

globuliferus
glob-yoo-LIH-fer-us
globulifera, globuliferum
With small spherical clusters, as
in *Saxifraga globulifera*

glutinosus
gloo-tin-OH-sus
glutinosa, glutinosum
Sticky; glutinous, as in *Eucryphia
glutinosa*

grandiceps
GRAN-dee-keps
With large head, as in *Leucogenes
grandiceps*

gummifer
GUM-mif-er
gummifera, gummiferum
Producing gum, as in *Seseli
gummiferum*

guttatus
goo-TAH-tus
guttata, guttatum
With spots, as in *Mimulus
guttatus*

🖙 Some plant names are
echoed in modern words. A
fissure is a long split and the
species name of the cleft-leaved
lady's mantle, *Alchemilla fissa*,
draws attention to the split
margin of the leaves.

heter-, hetero-

Used in compound words to denote various or diverse

hexa-

Used in compound words to denote six

hyalinus

hy-yuh-LEE-nus
hyalina, hyalinum
Transparent; almost transparent, as in *Allium hyalinum*

imbricans

im-brih-KANS

imbricatus

im-brih-KA-tus
imbricata, imbricatum
With elements that overlap in a regular pattern, as in *Gladiolus imbricatus*

inaequalis

in-ee-KWA-lis
inaequalis, inaequale
Unequal, as in *Geissorhiza inaequalis*

intumescens

in-tu-MES-enz
Swollen, as in *Carex intumescens*

irregularis

ir-reg-yoo-LAH-ris
irregularis, irregulare
With parts of different sizes, as in *Primula irregularis*

ladaniferus

lad-an-IH-fer-us
ladanifera, ladaniferum

ladanifer

lad-an-EE-fer
Producing ladanum, a fragrant gum resin, as in *Cistus ladanifer*

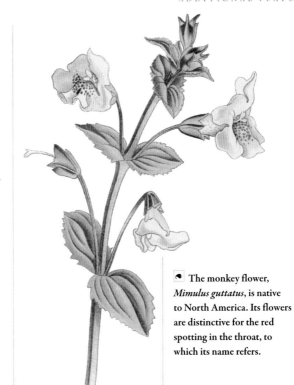

The monkey flower, **Mimulus guttatus**, is native to North America. Its flowers are distinctive for the red spotting in the throat, to which its name refers.

lati-

Used in compound words to denote broad

lept-

Used in compound words to denote thin or slender

lobatus

low-BAH-tus
lobata, lobatum
With lobes, as in *Cyananthus lobatus*

lobularis

lobe-yoo-LAY-ris
lobularis, lobulare
With lobes, as in *Narcissus lobularis*

longicuspis

lon-jih-KUS-pis
longicuspis, longicuspe
With long points, as in *Rosa longicuspis*

marginalis

mar-gin-AH-lis
marginalis, marginale

marginatus

mar-gin-AH-tus
marginata, marginatum
With a margin, sometimes used of variegated plants, as in *Saxifraga marginata*

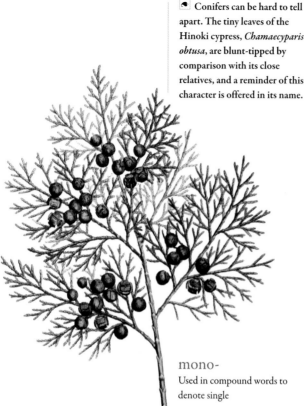

⟨🍃⟩ Conifers can be hard to tell apart. The tiny leaves of the Hinoki cypress, *Chamaecyparis obtusa*, are blunt-tipped by comparison with its close relatives, and a reminder of this character is offered in its name.

mono-
Used in compound words to denote single

marmoratus
mar-mor-RAH-tus
marmorata, marmoratum
marmoreus
mar-MOH-ree-us
marmorea, marmoreum
Marbled; mottled, as in *Kalanchoe marmorata*

megacanthus
meg-uh-KAN-thus
megacantha, megacanthum
With big spines, as in *Opuntia megacantha*

mucronatus
muh-kron-AH-tus
mucronata, mucronatum
With a point, as in *Gaultheria mucronata*

multiplex
MUL-tih-pleks
With many folds, as in *Bambusa multiplex*

muticus
MU-tih-kus
mutica, muticum
Blunt, as in *Pycnanthemum muticum*

mutilatus
mew-til-AH-tus
mutilata, mutilatum
Divided as though by tearing, as in *Peperomia mutilata*

nitens
NI-tenz
nitidus
NI-ti-dus
nitida, nitidum
Shining, as in *Lonicera nitida*

nudatus
noo-DAH-tus
nudata, nudatum
nudus
NEW-dus
nuda, nudum
Bare; naked, as in *Nepeta nuda*

obtusus
ob-TOO-sus
obtusa, obtusum
Blunt, as in *Chamaecyparis obtusa*

pachy-
Used in compound words to denote thick

pauci-
Used in compound words to denote few

penta-
Used in compound words to denote five

poly-
Used in compound words to denote many

ponderosus
pon-der-OH-sus
ponderosa, ponderosum
Heavy, as in *Pinus ponderosa*

punctatus
punk-TAH-tus
punctata, punctatum
With spots, as in *Anthemis punctata*

pungens
PUN-genz
With a sharp point, as in *Elymus pungens*

quadratus
kwad-RAH-tus
quadrata, quadratum
In fours, as in *Restio quadratus*

quinatus
kwi-NAH-tus
quinata, quinatum
In fives, as in *Akebia quinata*

radiatus
rad-ee-AH-tus
radiata, radiatum
With rays, as in *Pinus radiata*

reclinatus
rek-lin-AH-tus
reclinata, reclinatum
Bent backward, as in *Phoenix reclinata*

recurvatus
rek-er-VAH-tus
recurvata, recurvatum

recurvus
re-KUR-vus
recurva, recurvum
Curved backward, as in *Beaucarnea recurvata*

Many members of the cactus genus *Opuntia*, prickly pears, are surprisingly hardy in the US if given a well-drained soil. *O. megacantha*, however, comes with a warning; its name means "with big spines."

reflexus
ree-FLEKS-us
reflexa, reflexum

refractus
ray-FRAK-tus
refracta, refractum
Bent sharply backward, as in *Correa reflexa*

repandus
REP-an-dus
repanda, repandum
With wavy margins, as in *Cyclamen repandum*

resiniferus
res-in-IH-fer-us
resinifera, resiniferum

resinosus
res-in-OH-sus
resinosa, resinosum
Producing resin, as in *Euphorbia resinifera*

reticulatus
reh-tick-yoo-LAH-tus
reticulata, reticulatum
Netted, as in *Iris reticulata*

retortus
re-TOR-tus
retorta, retortum

retroflexus
ret-roh-FLEKS-us
retroflexa, retroflexum

retrofractus
re-troh-FRAK-tus
retrofracta, retrofractum
Twisted or turned backward, as in *Helichrysum retortum*

retusus
re-TOO-sus
retusa, retusum
With a rounded and notched tip, as in *Coryphantha retusa*

secundatus
see-kun-DAH-tus
secundata, secundatum
secundiflorus
sek-und-ee-FLOR-us
secundiflora,
secundiflorum
secundus
se-KUN-dus
secunda, secundum
With leaves or flowers growing
on one side of a stalk only,
as in *Echeveria secunda*

septemfidus
sep-TEM-fee-dus
septemfida, septemfidum
With seven divisions, as in
Gentiana septemfida

septemlobus
sep-tem-LOH-bus
septemloba, septemlobum
With seven lobes, as in *Primula
septemloba*

sinuatus
sin-yoo-AH-tus
sinuata, sinuatum
With a wavy margin, as in
Salpiglossis sinuata

squarrosus
skwa-ROH-sus
squarrosa, squarrosum
With spreading or curving parts
at the extremities, as in *Dicksonia
squarrosa*

steno-
Used in compound words to
denote narrow

**The Japanese
umbrella pine,
Sciadopitys verticillata,
is the last of an ancient
lineage of conifers that
makes a very attractive,
if slow-growing,
garden plant. Its name
perfectly describes
the picturesque
arrangement of its
thick leaves.**

striatus
stree-AH-tus
striata, striatum
With stripes, as in *Bletilla striata*

styracifluus
sty-rak-IF-lu-us
styraciflua, styracifluum
Producing gum, from *styrax*, the
Greek name for storax, as in
Liquidambar styraciflua

succulentus
suk-yoo-LEN-tus
succulenta, succulentum
Fleshy; juicy, as in *Oxalis
succulenta*

sulcatus
sul-KAH-tus
sulcata, sulcatum
With furrows, as in *Rubus
sulcatus*

tessellatus
tess-ell-AH-tus
tessellata, tessellatum
Chequered, as in *Indocalamus
tessellatus*

tetragonus
tet-ra-GON-us
tetragona, tetragonum
With four angles, as in
Nymphaea tetragona

tri-
Used in compound words to
denote three

triangularis
try-an-gew-LAH-ris
triangularis, triangulare
triangulatus
try-an-gew-LAIR-tus
triangulata, triangulatum
With three angles, as in *Oxalis
triangularis*

tricho-
Used in compound words to
denote hairy

trifasciata
try-fask-ee-AH-tuh
Three groups or bundles, as in
Sansevieria trifasciata

trifidus
TRY-fee-dus
trifida, trifidum
Cut in three, as in *Carex trifida*

trifurcatus
try-fur-KAH-tus
trifurcata, trifurcatum
With three forks, as in *Artemisia trifurcata*

tripartitus
try-par-TEE-tus
tripartita, tripartitum
With three parts, as in *Eryngium* × *tripartitum*

tripteris
TRIPT-er-is
tripterus
TRIPT-er-us
triptera, tripterum
With three wings, as in *Coreopsis tripteris*

truncatus
trunk-AH-tus
truncata, truncatum
Cut square, as in *Haworthia truncata*

uncinatus
un-sin-NA-tus
uncinata, uncinatum
With a hooked end, as in
Uncinia uncinata

uni-
Used in compound words to
denote one

verruculosus
ver-roo-ko-LOH-sus
verruculosa, verruculosum
With small warts, as in *Berberis verruculosa*

verticillatus
ver-ti-si-LAH-tus
verticillata, verticillatum
With a whorl or whorls, as in
Sciadopitys verticillata

vesicarius
ves-ee-KAH-ree-us
vesicaria, vesicarium
vesiculosus
ves-ee-kew-LOH-sus
vesiculosa, vesiculosum
Like a bladder; with small
bladders, as in *Eruca vesicaria*

vestitus
ves-TEE-tus
vestita, vestitum
Covered; clothed, as in *Sorbus vestita*

vittatus
vy-TAH-tus
vittata, vittatum
With lengthwise stripes, as in
Billbergia vittata

☛ The name *vittata* means
"with lengthwise stripes," so its
use for banded billbergia,
Billbergia vittata, may be
considered a misnomer as its
conspicuous markings run
across the leaves rather than
along them.

71

CHAPTER FOUR

Comparisons

An obvious way to explain something unknown
is by comparing it to something familiar, and a surprising
range of unlikely objects have been used in plant
names to illustrate an aspect of their appearance.
References to ostrich wings (*struthiopteris*), spider's webs
(*arachnoideum*), and candlesticks (*candelabrum*) can all
be found in the names of garden plants, each perfectly
capturing the character of some feature. Sometimes the
most evocative names of all, however, are those that allow
us to picture plants by likening them to other plants.

Animals

As naturalists, it is perhaps unsurprising that, when looking for ways to characterize plants, botanists' minds should turn to the animal kingdom. Through their names, some plants have been compared to parrots (*psittacina*), butterflies (*papilionaceum*), and zebras (*zebrina*), calling to mind a vivid picture of their qualities. Such analogies add an exoticism and color to plant names and understanding them brings an extra pleasure to gardening.

apiferus
a-PIH-fer-us
apifera, apiferum
Bearing bees, as in *Ophrys apifera*

aquilinus
ak-will-LEE-nus
aquilina, aquilinum
Like an eagle; aquiline, as in *Pteridium aquilinum*

coralloides
kor-al-OY-deez
Resembling coral, as in *Ozothamnus coralloides*

corniculatus
korn-ee-ku-LAH-tus
corniculata, corniculatum
With small horns, as in *Lotus corniculatus*

cornutus
kor-NOO-tus
cornuta, cornutum
With horns or shaped like a horn, as in *Viola cornuta*

◄ The dog's tooth violet, *Erythronium dens-canis*, gains both its common and scientific name from a fanciful likening of its long bulbs to the tooth of a dog. However, such comparisons are a helpful aide-mémoire.

crus-galli
krus GAL-ee
Cock's spur, as in *Crataegus crus-galli*

dens-canis
denz KAN-is
Term for dog's tooth, as in *Erythronium dens-canis*

dracunculus
dra-KUN-kyoo-lus
Small dragon, as in *Artemisia dracunculus*

echinatus
ek-in-AH-tus
echinata, echinatum
With spines like a hedgehog, as in *Pelargonium echinatum*

elephantipes
ell-uh-fan-TY-peez
Resembling an elephant's foot, as in *Yucca elephantipes*

erinaceus
er-in-uh-SEE-us
erinacea, erinaceum
Like a hedgehog, as in *Dianthus erinaceus*

hircinus
her-SEE-nus
hircina, hircinum
Goat-like; with a goat-like odor, as in *Hypericum hircinum*

hystrix
HIS-triks
Bristly; like a porcupine, as in *Colletia hystrix*

leonurus
lee-ON-or-us
leonura, leonurum
Like a lion's tail, as in *Leonotis leonurus*

leopardinus
leh-par-DEE-nus
leopardina, leopardinum
With spots like a leopard, as in *Calathea leopardina*

meleagris
mel-EE-uh-gris
meleagris, meleagre
With spots like a guinea fowl, as in *Fritillaria meleagris*

papilio
pap-ILL-ee-oh
A butterfly, as in *Hippeastrum papilio*

papilionaceus
pap-il-ee-on-uh-SEE-us
papilionacea, papilionaceum
Like a butterfly, as in *Pelargonium papilionaceum*

pardalinus
par-da-LEE-nus
pardalina, pardalinum

pardinus
par-DEE-nus
pardina, pardinum
With spots like a leopard, as in *Hippeastrum pardinum*

pedatifidus
ped-at-ee-FEE-dus
pedatifida, pedatifidum
Divided like a bird's foot, as in *Viola pedatifida*

pedatus
ped-AH-tus
pedata, pedatum
Shaped like a bird's foot, often in reference to the shape of a palmate leaf, as in *Adiantum pedatum*

penna-marina
PEN-uh mar-EE-nuh
Sea feather, as in *Blechnum penna-marina*

pinnatifidus
pin-nat-ih-FY-dus
pinnatifida, pinnatifidum
Cut in the form of a feather, as in *Eranthis pinnatifida*

pinnatus
pin-NAH-tus
pinnata, pinnatum
With leaves that grow from each side of a stalk; like a feather, as in *Santolina pinnata*

Snake's head fritillary, *Fritillaria meleagris,* embodies two animal comparisons. Its scientific name likens the chequered pattern of its flowers to the plumage of the guinea fowl, while its common name speaks for itself.

plumarius
ploo-MAH-ree-us
plumaria, plumarium
With feathers, as in *Dianthus plumarius*

plumosus
plum-OH-sus
plumosa, plumosum
Feathery, as in *Libocedrus plumosa*

● The scientific name of hart's tongue, *Asplenium scolopendrium*, was inspired by a perceived resemblance of the spore-bearing sori to centipede legs. Its common name likens the shape of its fronds to a stag's tongue.

testudinarius
tes-tuh-din-AIR-ee-us
testudinaria, testudinarium
Shaped like a turtle shell, as in *Durio testudinarius*

tigrinus
tig-REE-nus
tigrina, tigrinum
With stripes like the Asiatic tiger; with spots like a jaguar (known as "tiger" in South America), as in *Faucaria tigrina*

tragophylla
tra-go-FIL-uh
Literally goat leaf, as in *Lonicera tragophylla*

unguicularis
un-gwee-kew-LAH-ris
unguicularis, unguiculare

unguiculatus
un-gwee-kew-LAH-tus
unguiculata, unguiculatum
With claws, as in *Iris unguicularis*

urophyllus
ur-oh-FIL-us
urophylla, urophyllum
With leaves with a tip like a tail, as in *Clematis urophylla*

zebrinus
zeb-REE-nus
zebrina, zebrinum
With stripes like a zebra, as in *Tradescantia zebrina*

proboscideus
pro-bosk-ee-DEE-us
proboscidea, proboscideum
Shaped like a snout, as in *Arisarum proboscideum*

psittacinus
sit-uh-SIGN-us
psittacina, psittacinum

psittacorum
sit-a-KOR-um
Like a parrot; relating to parrots, as in V*riesea psittacinum*

rostratus
ro-STRAH-tus
rostrata, rostratum
With a beak, as in *Magnolia rostrata*

scolopendrius
skol-oh-PEND-ree-us
scolopendria, scolopendrium
From a supposed likeness of the underside of its fronds to a millipede or centipede (Greek *skolopendra*), as in *Asplenium scolopendrium*

scorpioides
skor-pee-OY-deez
Resembling a scorpion's tail, as in *Myosotis scorpioides*

struthiopteris
struth-ee-OP-ter-is
Like an ostrich wing, as in *Matteuccia struthiopteris*

Lotus corniculatus

Arisarum proboscideum

Dianthus plumarius

Animals in plants

The three names here are associated with animal body parts. Bird's foot trefoil, *Lotus corniculatus,* has fruit with little, horn-like tips. The spreading pods also provide the common name. The drawn-out flowers of the mousetail plant, *Arisarum proboscideum*, give it its name, meaning "shaped like a snout." Its common name may be more evocative. The crimped petals of the pink, *Dianthus plumarius*, are referred to in its epithet, meaning "with feathers."

Man-made objects

Human beings have fashioned plants into an extraordinary array of tools, ornaments, and other objects. In return, man-made objects have sometimes been adopted when giving names to plants in order to convey an idea of their natural appearance. References to weaponry are made quite frequently in names such as *clavatus* (club), *dolobratus* (hatchet), and *gladiatus* (sword), adding a combative edge to the gentle art of gardening.

acerosus
a-seh-ROH-sus
acerosa, acerosum
Like a needle, as in *Melaleuca acerosa*

acicularis
ass-ik-yew-LAH-ris
acicularis, aciculare
Shaped like a needle, as in *Rosa acicularis*

candelabrum
kan-del-AH-brum
Branched like a candelabra, as in *Salvia candelabrum*

clavatus
KLAV-ah-tus
clavata, clavatum
Shaped like a club, as in *Calochortus clavatus*

◖ The scientific name of the clubhair mariposa lily, *Calochortus clavatus*, means "shaped like a club" but this refers not to any large feature but to the bulbous-headed hairs that surround the nectary.

clypeatus
klye-pee-AH-tus
clypeata, clypeatum
Like a round Roman shield, as in *Fibigia clypeata*

coriarius
kor-i-AH-ree-us
coriaria, coriarium
Like leather, as in *Caesalpinia coriaria*

coronans
kor-OH-nanz

coronatus
kor-oh-NAH-tus
coronata, coronatum
Crowned, as in *Lychnis coronata*

cotyledon
kot-EE-lee-don
Small cup (referring to the leaves), as in *Lewisia cotyledon*

cucullatus
kuk-yoo-LAH-tus
cucullata, cucullatum
Like a hood, as in *Viola cucullata*

dolabratus
dol-uh-BRAH-tus
dolabrata, dolabratum

dolabriformis
doh-la-brih-FOR-mis
dolabriformis, dolabriforme
Shaped like a hatchet, as in *Thujopsis dolabrata*

ensatus
en-SA-tus
ensata, ensatum
In the shape of a sword, as in
Iris ensata

ensifolius
en-see-FOH-lee-us
ensifolia, ensifolium
With leaves shaped like a sword,
as in *Kniphofia ensifolia*

falcatus
fal-KAH-tus
falcata, falcatum
Shaped like a sickle, as in
Cyrtanthus falcatus

fenestralis
fen-ESS-tra-lis
fenestralis, fenestrale
With openings like a window, as
in *Vriesea fenestralis*

flagellaris
fla-gel-AH-ris
flagellaris, flagellare

flagelliformis
fla-gel-ih-FOR-mis
flagelliformis,
flagelliforme
Like a whip; with long, thin
shoots, as in *Celastrus flagellaris*

galeatus
ga-le-AH-tus
galeata, galeatum

galericulatus
gal-er-ee-koo-LAH-tus
galericulata, galericulatum
Shaped like a helmet, as in
Sparaxis galeata

◄ The Japanese flag iris, *Iris
ensata*, is a delightful plant for
boggy soil and bears large,
flamboyant, purple or red
flowers. However, it is for its
sword-shaped leaves that it
received its name.

gemmatus
jem-AH-tus
gemmata, gemmatum
Bejewelled, as in *Wikstroemia
gemmata*

gladiatus
glad-ee-AH-tus
gladiata, gladiatum
Like a sword, as in *Coreopsis
gladiata*

hastatus
hass-TAH-tus
hastata, hastatum
Shaped like a spear, as in *Verbena
hastata*

infundibuliformis
in-fun-dih-bew-LEE-for-mis
infundibuliformis,
infundibuliforme
In the shape of a funnel or
trumpet, as in *Crossandra
infundibuliformis*

lanceolatus
lan-see-oh-LAH-tus
lanceolata, lanceolatum

lanceus
lan-SEE-us
lancea, lanceum
In the shape of a spear, as in
Drimys lanceolata

manicatus
mah-nuh-KAH-tus
manicata, manicatum
With long sleeves, as in *Gunnera
manicata*

martagon
MART-uh-gon
A word of uncertain origin,
thought in *Lilium martagon* to
refer to the turban-like flower

nummularius
num-ew-LAH-ree-us
nummularia,
nummularium
Like coins, as in *Lysimachia
nummularia*

papyraceus
pap-ih-REE-see-us
papyracea, papyraceum
Like paper, as in *Narcissus papyraceus*

pectinatus
pek-tin-AH-tus
pectinata, pectinatum
Like a comb, as in *Euryops pectinatus*

peltatus
pel-TAH-tus
peltata, peltatum
Shaped like a shield, as in *Darmera peltata*

pulvinatus
pul-vin-AH-tus
pulvinata, pulvinatum
Like a cushion, as in *Echeveria pulvinata*

pyramidalis
peer-uh-mid-AH-lis
pyramidalis, pyramidale
Shaped like a pyramid, as in *Ornithogalum pyramidale*

saccatus
sak-KAH-tus
saccata, saccatum
Like a bag, or saccate, as in *Lonicera saccata*

sceptrum
SEP-trum
Like a scepter, as in *Digitalis sceptrum*

scoparius
sko-PAIR-ee-us
scoparia, scoparium
Like broom, as in *Cytisus scoparius*

scutatus
skut-AH-tus
scutata, scutatum

scutellaris
skew-tel-AH-ris
scutellaris, scutellare

scutellatus
skew-tel-LAH-tus
scutellata, scutellatum
Shaped like a shield or platter, as in *Rumex scutatus*

strumosus
stroo-MOH-sus
strumosa, strumosum
With cushion-like swellings, as in *Nemesia strumosa*

subulatus
sub-yoo-LAH-tus
subulata, subulatum
Awl- or needle-shaped, as in *Phlox subulata*

tazetta
taz-ET-tuh
Little cup, as in *Narcissus tazetta*

thyrsoideus
thurs-OY-dee-us
thyrsoidea, thyrsoideum

thyrsoides
thurs-OY-deez
Like a Bacchic staff, as in *Ornithogalum thyrsoides*

urceolatus
ur-kee-oh-LAH-tus
urceolata, urceolatum
Shaped like an urn, as in *Galax urceolata*

velutinus
vel-oo-TEE-nus
velutina, velutinum
Like velvet, as in *Musa velutina*

French sorrel, *Rumex scutatus*, is sometimes used as a slightly bitter cooking herb. Another common name, shield leaf sorrel, reflects its scientific name, which refers to the shape of the leaves.

Madeiran foxglove, *Digitalis sceptrum*, is a shrubby relative of the common foxglove. Its leafless, scepter-like inflorescence with flowers borne all around the stem is one of the things that differentiate it from its more familiar cousins.

Natural objects

A curious miscellany of substances, from nests (*nidus*) to dust (*pulverulentus*), have made their way into the names given to our garden plants. The rationale behind this may not be obvious until it is understood that the first of these refers to the characteristic shape of the bird's-nest fern, *Asplenium nidus*, and the second to the powdery leaves and stems of mealy primrose, *Primula pulverulenta*.

arachnoides
a-rak-NOY-deez
arachnoideus
a-rak-NOY-dee-us
arachnoidea,
arachnoideum
Like a spider's web, as in
Sempervivum arachnoideum

auriculatus
aw-rik-yoo-LAH-tus
auriculata, auriculatum
auriculus
aw-RIK-yoo-lus
auricula, auriculum
auritus
aw-RY-tus
aurita, auritum
With ears or ear-shaped
appendages, as in *Plumbago
auriculata*

blepharophyllus
blef-ar-oh-FIL-us
blepharophylla,
blepharophyllum
With leaves that are fringed like
eyelashes, as in *Arabis
blepharophylla*

botryoides
bot-ROY-deez
Resembling a bunch of grapes, as
in *Muscari botryoides*

▶ The Venus maidenhair,
Adiantum capillus-veneris,
is a delicate and elegant
fern found in damp
habitats throughout much
of the world. Its name
romantically refers to a
likeness of its cascading
fronds to the hair of Venus.

calycinus
ka-lih-KEE-nus
calycina, calycinum
Like a calyx, as in *Halimium
calycinum*

capillaris
kap-ill-AH-ris
capillaris, capillare
Very slender, like fine hair, as in
Tillandsia capillaris

capillus-veneris
KAP-il-is VEN-er-is
Venus's hair, as in *Adiantum
capillus-veneris*

caudatus
kaw-DAH-tus
caudata, caudatum
With a tail, as in *Asarum
caudatum*

cheiri
kye-EE-ee
Perhaps from the Greek word
cheir, hand, as in *Erysimum
cheiri*

ciliosus
sil-ee-OH-sus
ciliosa, ciliosum
With a small fringe, as in
Sempervivum ciliosum

dactyliferus
dak-ty-LIH-fer-us
dactylifera, dactyliferum
With fingers; finger-like, as in
Phoenix dactylifera

digitalis
dij-ee-TAH-lis
digitalis, digitale
Like a finger, as in *Penstemon
digitalis*

digitatus
dig-ee-TAH-tus
digitata, digitatum
Like the shape of an open hand,
as in *Schefflera digitata*

granulatus
gran-yoo-LAH-tus
granulata, granulatum
Bearing grain-like structures, as
in *Saxifraga granulata*

hypoglottis
hy-poh-GLOT-tis
Underside of the tongue, from
the shape of the pods, as in
Astragalus hypoglottis

lentiginosus
len-tig-ih-NOH-sus
lentiginosa, lentiginosum
With freckles, as in *Coelogyne
lentiginosa*

lingua
LIN-gwa
Tongue; like a tongue, as in
Pyrrosia lingua

The origin of the scientific name of the common wallflower, *Erysimum cheiri*, is not certain, but one pleasing notion is that it is from the Greek word *cheri*, hand, referring to the carrying of these flowers as bouquets.

lunatus
loo-NAH-tus
lunata, lunatum

lunulatus
loo-nu-LAH-tus
lunulata, lunulatum
Shaped like the crescent moon,
as in *Cyathea lunulata*

macrobotrys
mak-ro-BOT-rees
With large, grape-like clusters, as
in *Strongylodon macrobotrys*

mammillatus
mam-mil-LAIR-tus
mammillata,
mammillatum

mammillaris
mam-mil-LAH-ris
mammillaris, mammillare

mammosus
mam-OH-sus
mammosa, mammosum
Bearing nipple- or breast-like
structures, as in *Solanum
mammosum*

Tulipa clusiana var. *stellata*

BEHIND THE NAME

Stardust

The names of the two plants here together make stardust. The epithet *stellata* means starry, and in this variety of the lady tulip, *Tulipa clusiana* var. *stellata* refers to the shape of the open flowers. Mealy primrose, *Primula pulverulenta*, is an attractive perennial from China. It owes its name, which means "appearing to be covered in dust," to the grainy, white layer (farina) that covers the stems.

Primula pulverulenta

membranaceus
mem-bran-AY-see-us
membranacea,
membranaceum
Like skin or membrane, as in
Scadoxus membranaceus

nidus
NID-us
Like a nest, as in *Asplenium nidus*

oculatus
ock-yoo-LAH-tus
oculata, oculatum
With an eye, as in *Haworthia oculata*

pisiferus
pih-SIH-fer-us
pisifera, pisiferum
Like peas, as in *Chamaecyparis pisifera*

polyblepharus
pol-ee-BLEF-ar-us
polyblephara,
polyblepharum
With many fringes or eyelashes,
as in *Polystichum polyblepharum*

pomiferus
pom-IH-fer-us
pomifera, pomiferum
Like apples, as in *Maclura pomifera*

pruinatus
proo-in-AH-tus
pruinata, pruinatum

pruinosus
proo-in-NOH-sus
pruinosa, pruinosum
Glistening like frost, as in
Cotoneaster pruinosus

pulverulentus
pul-ver-oo-LEN-tus
pulverulenta,
pulverulentum
Appearing to be covered in dust,
as in *Primula pulverulenta*

pustulatus
pus-tew-LAH-tus
pustulata, pustulatum
Appearing to be blistered, as in
Lachenalia pustulata

reniformis
ren-ih-FOR-mis
reniformis, reniforme
Shaped like a kidney, as in
Begonia reniformis

sideroxylon
sy-der-oh-ZY-lon
Iron-like wood, as in *Eucalyptus sideroxylon*

stellaris
stell-AH-ris
stellaris, stellare

stellatus
stell-AH-tus
stellata, stellatum
Starry, as in *Magnolia stellata*

uva-crispa
OO-vuh KRIS-puh
Curled grape, as in *Ribes uva-crispa*

uvaria
oo-VAR-ee-uh
Like a bunch of grapes, as in
Kniphofia uvaria

The similarity to grapes implied by the scientific name of the gooseberry, *Ribes uva-crispa*, is understandable enough but the curly aspect is less clear. This probably refers to the frizzy hairs that cover the fruits.

Appearance of other plants

Convergent evolution helps to explain similarities in the appearance of creatures from different parts of the tree of life. Bats and sparrows both have wings but the first is a mammal and the second a bird. They have adapted to meet life's needs in similar ways, and so it is with some plants. The wheel tree, *Trochodendron aralioides*, is only distantly related to *Aralia* but something of its appearance can be predicted from its name.

abies
A-bees
abietinus
ay-bee-TEE-nus
abietina, abietinum
Like a fir tree, *Abies*, as in *Picea abies*

agavoides
ah-gav-OY-deez
Resembling *Agave*, as in *Echeveria agavoides*

ageratoides
ad-jur-rat-OY-deez
Resembling *Ageratum*, as in *Aster trifoliatus* subsp. *ageratoides*

aizoides
ay-ZOY-deez
Like the genus *Aizoon*, as in *Saxifraga aizoides*

alliaceus
al-lee-AY-see-us
alliacea, alliacum
Like onion or garlic, *Allium*, as in *Tulbaghia alliacea*

◀ The diverse saxifrage genus, *Saxifraga,* numbers more than 400 species, mostly from mountainous areas of the world. The yellow mountain saxifrage, *Saxifraga aizoides,* resembles a member of the much smaller genus *Aizoon.*

alopecuroides
al-oh-pek-yur-OY-deez
Like the foxtail genus, *Alopecurus,* as in *Pennisetum alopecuroides*

amelloides
am-el-OY-deez
Resembling *Aster amellus* (from its Roman name), as in *Felicia amelloides*

amygdaloides
am-ig-duh-LOY-deez
Resembling almond, as in *Euphorbia amygdaloides*

anthemoides
an-them-OY-deez
Resembling chamomile (Greek *anthemis*), as in *Rhodanthe anthemoides*

aralioides
a-ray-lee-OY-deez
Like *Aralia*, as in *Trochodendron aralioides*

artemisioides
ar-tem-iss-ee-OY-deez
Resembling *Artemisia*, as in *Senna artemisioides*

arundinaceus
a-run-din-uh-KEE-us
arundinacea, arundinaceum
Like a reed, as in *Phalaris arundinacea*

asclepiadeus
ass-cle-pee-AD-ee-us
asclepiadea, asclepiadeum
Like milkweed, *Asclepias*, as in
Gentiana asclepiadea

asparagoides
as-par-a-GOY-deez
Resembling or like asparagus, as
in *Acacia asparagoides*

asphodeloides
ass-fo-del-OY-deez
Like *Asphodelus*, as in *Geranium
asphodeloides*

asteroides
ass-ter-OY-deez
Resembling *Aster*, as in *Amellus
asteroides*

astilboides
a-stil-BOY-deez
Resembling *Astilbe*, as in *Astilbe
astilboides*

aubrietioides
au-bre-teh-OY-deez
Resembling *Aubrieta*, as in
Arabis aubrietiodes

bambusoides
bam-BOO-soy-deez
Resembling bamboo, *Bambusa*,
as in *Phyllostachys bambusoides*

bellidiformis
bel-id-EE-for-mis
bellidiformis, bellidiforme
Like a daisy, *Bellis*, as in
Dorotheanthus bellidiformis

■ The wheel tree,
*Trochodendron
aralioides*, bears a
remarkable similarity
to a member of the
Aralia family.

betaceus
bet-uh-KEE-us
betacea, betaceum
Like a beet, *Beta*, as in *Solanum
betaceum*

betulinus
bet-yoo-LEE-nus
betulina, betulinum

betuloides
bet-yoo-LOY-deez
Resembling or like a birch,
Betula, as in *Carpinus betulinus*

bignonioides
big-non-YOY-deez
Resembling crossvine, *Bignonia*,
as in *Catalpa bignonioides*

bromoides
brom-OY-deez
Resembling brome grass,
Bromus, as in *Stipa bromoides*

bryoides
bri-ROY-deez
Resembling moss, as in *Dionysia
bryoides*

cannabinus
kan-na-BEE-nus
cannabina, cannabinum
Like hemp, *Cannabis*, as in
Eupatorium cannabinum

cardunculus
kar-DUNK-yoo-lus
carduncula, cardunculum
Like a small thistle, as in *Cynara
cardunculus*

cautleyoides
kawt-ley-OY-deez
Resembling *Cautleya*, as in
Roscoea cautleyoides

⬧ **Some species of the genus *Erodium* have very cut leaves, while common stork's bill, *E. cicutarium*, resembles hemlock, *Conium maculatum* (previously *Cicuta*), one of the umbel family.**

chamaebuxus
kam-ay-BUKS-us
Dwarf boxwood, *Buxus*, as in *Polygala chamaebuxus*

chamaecyparissus
kam-ee-ky-pah-RIS-us
Like *Chamaecyparis*, as in *Santolina chamaecyparissus*

cicutarius
kik-u-TAH-ree-us
cicutaria, cicutarium
Like water hemlock, *Conium maculatum* (formerly *Cicuta*), as in *Erodium cicutarium*

citratus
sit-TRAH-tus
citrata, citratum
Like *Citrus*, as in *Aerangis citrata*

clethroides
klee-THROY-deez
Resembling white alder, *Clethra*, as in *Lysimachia clethroides*

cneorum
suh-NOR-um
From the Greek for a small, olive-like shrub, possibly a kind of *Daphne*, as in *Convolvulus cneorum*

convallarioides
kon-va-lar-ee-OY-deez
Resembling lily-of-the-valley, *Convallaria*, as in *Speirantha convallarioides*

cortusoides
kor-too-SOY-deez
Resembling *Cortusa*, as in *Primula cortusoides*

crocosmiiflorus
kroh-koz-mee-eye-FLOR-us
crocosmiiflora, crocosmiiflorum
With flowers like *Crocosmia*. *Crocosmia × crocosmiiflora* was originally in the genus *Montbretia*; the name therefore meant crocosmia-flowered montbretia

cupressinus
koo-pres-EE-nus
cupressina, cupressinum

cupressoides
koo-press-OY-deez
Resembling cypress, *Cupressus*, as in *Fitzroya cupressoides*

cyclamineus
SIGH-kluh-min-ee-us
cyclaminea, cyclamineum
Like *Cyclamen*, as in *Narcissus cyclamineus*

cynaroides
sin-nar-OY-deez
Resembling *Cynara*, as in *Protea cynaroides*

daphnoides
daf-NOY-deez
Resembling *Daphne*, as in *Salix daphnoides*

daucoides
do-KOY-deez
Resembling carrot, *Daucus*, as in *Erodium daucoides*

dipsaceus
dip-SAK-ee-us
dipsacea, dipsaceum
Like teasel, *Dipsacus*, as in *Carex dipsacea*

echioides
ek-ee-OY-deez
Resembling viper's bugloss,
Echium, as in *Picris echioides*

ericoides
er-ik-OY-deez
Resembling heather, *Erica*, as in
Symphyotrichum ericoides

eugenioides
yoo-jee-nee-OY-deez
Resembling the genus *Eugenia*,
as in *Pittosporum eugenioides*

ficoides
fy-KOY-deez

ficoideus
fy-KOY-dee-us
ficoidea, ficoideum
Resembling a fig, *Ficus*, as in
Senecio ficoides

filipendulus
fil-ih-PEN-dyoo-lus
filipendula, filipendulum
Like meadow-sweet, *Filipendula*,
as in *Oenanthe filipendula*

foeniculaceus
fen-ee-kul-ah-KEE-us
foeniculacea,
foeniculaceum
Like fennel, *Foeniculum*, as in
Argyranthemum foeniculaceum

fragarioides
fray-gare-ee-OY-deez
Resembling strawberry, *Fragaria*,
as in *Waldsteinia fragarioides*

geoides
jee-OY-deez
Like avens, *Geum*, as in
Waldsteinia geoides

gossypinus
goss-ee-PEE-nus
gossypina, gossypinum
Like cotton, *Gossypium*, as in
Strobilanthes gossypina

gramineus
gram-IN-ee-us
graminea, gramineum
Like grass, as in *Iris graminea*

hakeoides
hak-ee-OY-deez
Resembles *Hakea*, as in *Berberis
hakeoides*

hederaceus
hed-er-AYE-see-us
hederacea, hederaceum
Like ivy, *Hedera*, as in *Glechoma
hederacea*

helianthoides
hel-ih-anth-OH-deez
Resembling sunflower,
Helianthus, as in *Heliopsis
helianthoides*

hydrangeoides
hy-drain-jee-OY-deez
Resembling *Hydrangea*, as in
Schizophragma hydrangeoides

hypnoides
hip-NO-deez
Resembling moss, as in *Saxifraga
hypnoides*

intybaceus
in-tee-BAK-ee-us
intybacea, intybaceum
Like chicory, *Cichorium intybus*,
as in *Hieracium intybaceum*

iridiflorus
ir-id-uh-FLOR-us
iridiflora, iridiflorum
With flowers like *Iris*, as in
Canna iridiflora

✏ Nowadays, the grass family
is generally known by the name
Poaceae, but it was previously
called *Gramineae* and it is that
name that is referred to in the
epithet of the grass-leaved
iris, *Iris graminea*.

iridioides
ir-id-ee-OY-deez
Resembling *Iris*, as in *Dietes iridioides*

ixioides
iks-ee-OY-deez
Resembling corn lily, *Ixia*, as in *Libertia ixioides*

jasmineus
jaz-MIN-ee-us
jasminea, jasmineum
Like jasmine, *Jasminum*, as in *Daphne jasminea*

jasminoides
jaz-min-OY-deez
Resembling jasmine, *Jasminum*, as in *Trachelospermum jasminoides*

juniperinus
joo-nip-er-EE-nus
juniperina, juniperinum
Like juniper, *Juniperus*; blue-black, as in *Grevillea juniperina*

laricinus
lar-ih-SEE-nus
laricina, laricinum
Like a larch, *Larix*, as in *Banksia laricina*

ligustrinus
lig-us-TREE-nus
ligustrina, ligustrinum
Like privet, *Ligustrum*, as in *Ageratina ligustrina*

lili-
Used in compound words to denote lily

liliaceus
lil-lee-AY-see-us
liliacea, liliaceum
Like lily, *Lilium*, as in *Fritillaria liliacea*

linifolius
lin-ih-FOH-lee-us
linifolia, linifolium
With leaves like flax, *Linum*, as in *Tulipa linifolia*

lobelioides
lo-bell-ee-OH-id-ees
Resembling *Lobelia*, as in *Wahlenbergia lobelioides*

◀ The monk's hood, *Aconitum napellus*, is distinctive for its cut leaves and hooded, blue flowers. However, its name refers to neither of these features but to the resemblance of its roots to a little turnip.

loliaceus
loh-lee-uh-SEE-us
loliacea, loliaceum
Like rye-grass, *Lolium*, as in × *Festulolium loliaceum*

lupulinus
lup-oo-LEE-nus
lupulina, lupulinum
Like hop, *Humulus lupulus*, as in *Medicago lupulina*

lycopodioides
ly-kop-oh-dee-OY-deez
Resembling clubmoss, *Lycopodium*, as in *Cassiope lycopodioides*

malvaceus
mal-VAY-see-us
malvacea, malvaceum
Like mallow, *Malva*, as in *Physocarpus malvaceus*

miliaceus
mil-ee-AY-see-us
miliacea, miliaceum
Relating to millet, as in *Panicum miliaceum*

mimosoides
mim-yoo-SOY-deez
Resembling *Mimosa*, as in *Caesalpinia mimosoides*

muscoides
mus-COY-deez
Resembling moss, as in *Saxifraga muscoides*

myrsinites
mir-SIN-ih-teez

myrsinoides
mir-sy-NOY-deez
Resembling *Myrsine*, as in *Gaultheria myrsinoides*

napellus
nap-ELL-us
napella, napellum
Like a little turnip, referring to
the roots, as in *Aconitum
napellus*

nepetoides
nep-et-OY-deez
Resembling catmint, *Nepeta*,
as in *Agastache nepetoides*

ocymoides
ok-kye-MOY-deez
Resembling basil, *Ocimum*,
as in *Halimium ocymoides*

oleoides
oh-lee-OY-deez
Resembling olive, *Olea*, as in
Daphne oleoides

orchideus
or-KI-de-us
orchidea, orchideum

orchioides
or-ki-OY-deez
Like an orchid, *Orchis*, as in
Veronica orchidea

pimeleoides
py-mee-lee-OY-deez
Resembling *Pimelea*, as in
Pittosporum pimeleoides

pimpinellifolius
pim-pi-nel-ih-FOH-lee-us
pimpinellifolia,
pimpinellifolium
With leaves like anise,
Pimpinella, as in *Rosa
pimpinellifolia*

◖ The jaggedly
lobed leaves of the
Norway maple, *Acer
platanoides*, resemble
the London plane,
Platanus × hispanica.
This similarity is
reflected in its name.

pineus
PY-nee-us
pinea, pineum
Relating to pine, *Pinus*, as in
Pinus pinea

pininana
pin-in-AH-nuh
Dwarf pine, as in *Echium
pininana*

plantagineus
plan-tuh-JIN-ee-us
plantaginea,
plantagineum
Like plantain, *Platago*, as in
Hosta plantaginea

platanoides
pla-tan-OY-deez
Resembling a plane tree,
Platanus, as in *Acer platanoides*

plumbaginoides
plum-bah-gih-NOY-deez
Resembling *Plumbago*, as in
Ceratostigma plumbaginoides

polygonoides
pol-ee-gon-OY-deez
Resembling *Polygonum*, as in
Alternanthera polygonoides

populneus
pop-ULL-nee-us
populnea, populneum
Relating to poplar, *Populus*,
as in *Brachychiton populneus*

The sea buckthorn, *Hippophae rhamnoides*, thrives in poor soil and produces edible fruits commonly used to make juice, jams, or jellies. It's also an attractive landscape plant.

primulinus
prim-yoo-LEE-nus
primulina, primulinum

primuloides
prim-yoo-LOY-deez
Like primrose, *Primula*, as in
Paphiopedilum primulinum

pseudacorus
soo-DA-ko-rus
Deceptively like *Acorus*, as in *Iris pseudacorus*

pseudocamellia
soo-doh-kuh-MEE-lee-uh
Deceptively like *Camellia*, as in
Stewartia pseudocamellia

pseudodictamnus
soo-do-dik-TAM-nus
Deceptively like *Dictamnus*, as in
Ballota pseudodictamnus

pseudonarcissus
soo-doh-nar-SIS-us
Deceptively like *Narcissus*; in
N. pseudonarcissus the *Narcissus*
referred to is *N. poeticus*

pteridoides
ter-id-OY-deez
Resembling *Pteris*, as in *Coriaria pteridoides*

pulegioides
pul-eg-ee-OY-deez
Like pennyroyal, *Mentha pulegium*, as in *Thymus pulegioides*

ranunculoides
ra-nun-kul-OY-deez
Resembling buttercup,
Ranunculus, as in *Anemone ranunculoides*

rhamnoides
ram-NOY-deez
Resembling buckthorn,
Rhamnus, as in *Hippophae rhamnoides*

rosaceus
ro-ZAY-see-us
rosacea, rosaceum
Rose-like, as in *Saxifraga rosacea*

salicarius
sa-lih-KAH-ree-us
salicaria, salicarium
Like willow, *Salix*, as in *Lythrum salicaria*

salicinus
sah-lih-SEE-nus
salicina, salicinum
Like willow, *Salix*, as in *Prunus salicina*

salignus
sal-LIG-nus
saligna, salignum
Like willow, *Salix*, as in
Podocarpus salignus

schoenoprasum
skee-no-PRAY-zum
Epithet for chives, *Allium schoenoprasum*, meaning
"rush-leek" in Greek

scilloides
sil-OY-deez
Resembling *Scilla*, as in
Puschkinia scilloides

selaginoides
sel-ag-ee-NOY-deez
Resembling clubmoss,
Selaginella, as in *Athrotaxis
selaginoides*

thalictroides
thal-ik-TROY-deez
Resembling meadow rue,
Thalictrum, as in *Anemonella
thalictroides*

thymoides
ty-MOY-deez
Resembling thyme, *Thymus*, as
in *Eriogonum thymoides*

tiarelloides
tee-uh-rell-OY-deez
Resembling *Tiarella*, as in
× *Heucherella tiarelloides*

tulipiferus
too-lip-IH-fer-us
tulipifera, tulipiferum
Producing tulips or tulip-like
flowers, as in *Liriodendron
tulipifera*

typhinus
ty-FEE-nus
typhina, typhinum
Like reedmace, *Typha*, as in
Rhus typhina

ulmaria
ul-MAR-ee-uh
Like elm, *Ulmus*, as in
Filipendula ulmaria

viburnoides
vy-burn-OY-deez
Resembling *Viburnum*, as in
Pileostegia viburnoides

vitaceus
vee-TAY-see-us
vitacea, vitaceum
Like vine, *Vitis*, as in
Parthenocissus vitacea

yuccoides
yuk-KOY-deez
Resembling *Yucca*, as in
Beschorneria yuccoides

❧ Purple loosestrife, *Lythrum
salicaria*, is a native plant in the
UK, where it is found in
marshy soils. In the US it is
highly invasive. Its name refers
to the similarity of its leaves to
another plant of damp habitats,
willow, *Salix*.

Leaves of other plants

Many gardeners will feel confident in identifying the broad, lobed leaves of the grape vine, *Vitis vinifera*, or the contrasting spiky needles of the common juniper, *Juniperus communis*. However, foliage can be a poor gauge of relationship and sometimes distantly related plants have strikingly similar leaves. Consequently we have the vine-leaved abutilon, *Abutilon vitifolium*, and the juniper-leaved thrift, *Armeria juniperifolia*.

abrotanifolius
ab-ro-tan-ih-FOH-lee-us
abrotanifolia,
abrotanifolium
With leaves like southernwood, *Artemisia abrotanum*, as in *Euryops abrotanifolius*

acanthifolius
a-kanth-ih-FOH-lee-us
acanthifolia,
acanthifolium
With leaves like *Acanthus*, as in *Carlina acanthifolia*

acerifolius
a-ser-ih-FOH-lee-us
acerifolia, acerifolium
With leaves like maple, *Acer*, as in *Quercus acerifolia*

◥ Many members of the buttercup family, *Ranunculaceae*, have lobed leaves, and these can be very similar. Fair maids of Kent, *Ranunculus aconitifolius*, has foliage that resembles monk's hood, *Aconitum*.

achilleifolius
ah-key-lee-FOH-lee-us
achilleifolia,
achilleifolium
With leaves like common yarrow, *Achillea millefolium*, as in *Tanacetum achilleifolium*

aconitifolius
a-kon-eye-tee-FOH-lee-us
aconitifolia, aconitifolium
With leaves like aconite, *Aconitum*, as in *Ranunculus aconitifolius*

adiantifolius
ad-ee-an-tee-FOH-lee-us
adiantifolia, adiantifolium
With leaves like maidenhair fern, *Adiantum*, as in *Anemia adiantifolia*

aesculifolius
es-kew-li-FOH-lee-us
aesculifolia, aesculifolium
With leaves like horse chestnut, *Aesculus*, as in *Rodgersia aesculifolia*

alliariifolius
al-ee-ar-ee-FOH-lee-us
alliariifolia, alliariifolium
With leaves like *Alliaria*, as in *Valeriana alliariifolia*

alnifolius
al-nee-FOH-lee-us
alnifolia, alnifolium
With leaves like alder, *Alnus*, as in *Sorbus alnifolia*

aquifolius
a-kwee-FOH-lee-us
aquifolia, aquifolium
Holly-leaved (from the Latin
name for holly, *aquifolium*), as in
Mahonia aquifolium

aquilegiifolius
ak-wil-egg-ee-FOH-lee-us
aquilegiifolia,
aquilegiifolium
With leaves like columbine,
Aquilegia, as in *Thalictrum
aquilegiifolium*

arbutifolius
ar-bew-tih-FOH-lee-us
arbutifolia, arbutifolium
With leaves like the strawberry
tree, *Arbutus*, as in *Aronia
arbutifolia*

atriplicifolius
at-ry-pliss-ih-FOH-lee-us
atriplicifolia,
atriplicifolium
With leaves like orache or
salt-bush, *Atriplex*, as in
Perovskia atriplicifolia

betonicifolius
bet-on-ih-see-FOH-lee-us
betonicifolia,
betonicifolium
Like betony, *Stachys*, as in
Meconopsis betonicifolia

buxifolius
buks-ih-FOH-lee-us
buxifolia, buxifolium
With leaves like box, *Buxus*, as in
Cantua buxifolia

 Red chokeberry, *Aronia
arbutifolia*, is a member of
the rose family from North
America. Its leaves, however,
are like those of the strawberry
tree, *Arbutus*.

carpinifolius
kar-pine-ih-FOH-lee-us
carpinifolia, carpinifolium
With leaves like hornbeam,
Carpinus, as in *Zelkova
carpinifolia*

caryophyllus
kar-ee-oh-FIL-us
caryophylla, caryophyllum
Walnut-leaved (from Greek
karya); likened to clove for their
smell, and thence to clove pink,
as in *Dianthus caryophyllus*

cercidifolius
ser-uh-sid-ih-FOH-lee-us
cercidifolia, cercidifolium
With leaves like redbud tree,
Cercis, as in *Disanthus
cercidifolius*

cinnamomifolius
sin-nuh-mom-ih-FOH-lee-us
cinnamomifolia,
cinnamomifolium
With leaves like cinnamon,
Cinnamomum, as in *Viburnum
cinnamomifolium*

cissifolius
kiss-ih-FOH-lee-us
cissifolia, cissifolium
With leaves like ivy (from the
Greek *kissos*), as in *Acer
cissifolium*

corylifolius
kor-ee-lee-FOH-lee-us
corylifolia, corylifolium
With leaves like hazelnut,
Corylus, as in *Betula corylifolia*

The trigger plant, *Stylidium graminifolium*, owes its common name to its flower parts, which are triggered by touch to cover visiting insects in pollen. However, its Latin name refers to its grass-like leaves.

cotinifolius
kot-in-ih-FOH-lee-us
cotinifolia, cotinifolium
With leaves like the smoke tree, *Cotinus*, as in *Euphorbia cotinifolia*

crataegifolius
krah-tee-gi-FOH-lee-us
crataegifolia, crataegifolium
With leaves like hawthorn, *Crataegus*, as in *Acer crataegifolium*

delphinifolius
del-fin-uh-FOH-lee-us
delphinifolia, delphinifolium
With leaves like *Delphinium*, as in *Aconitum delphinifolium*

elaeagnifolius
el-ee-ag-ne-FOH-lee-us
elaeagnifolia, elaeagnifolium
With leaves like *Elaeagnus*, as in *Brachyglottis elaeagnifolia*

empetrifolius
em-pet-rih-FOH-lee-us
empetrifolia, empetrifolium
With leaves like crowberry, *Empetrum*, as in *Berberis empetrifolia*

equisetifolius
ek-wih-set-ih-FOH-lee-us
equisetifolia, equisetifolium

equisetiformis
eck-kwiss-ee-tih-FOR-mis
equisetiformis, equisetiforme
Resembling horsetail, *Equisetum*, as in *Russelia equisetiformis*

fagifolius
fag-ih-FOH-lee-us
fagifolia, fagifolium
With leaves like beech, *Fagus*, as in *Clethra fagifolia*

filicifolius
fil-ee-kee-FOH-lee-us
filicifolia, filicifolium
With leaves like a fern, as in *Polyscias filicifolia*

fraxinifolius
fraks-in-ee-FOH-lee-us
fraxinifolia, fraxinifolium
With leaves like ash, *Fraxinus*, as in *Pterocarya fraxinifolia*

genistifolius
jih-nis-tih-FOH-lee-us
genistifolia, genistifolium
With leaves like broom, *Genista*, as in *Linaria genistifolia*

graminifolius
gram-in-ee-FOH-lee-us
graminifolia, graminifolium
With grass-like leaves, as in *Stylidium graminifolium*

hederifolius
hed-er-ih-FOH-lee-us
hederifolia, hederifolium
With leaves like ivy, *Hedera*, as in *Veronica hederifolia*

hepaticifolius
hep-at-ih-sih-FOH-lee-us
hepaticifolia, hepaticifolium
With leaves like liverwort, *Hepatica*, as in *Cymbalaria hepaticifolia*

heracleifolius
hair-uh-klee-ih-FOH-lee-us
heracleifolia, heracleifolium
With leaves like hogweed, *Heracleum*, as in *Begonia heracleifolia*

hypericifolius
hy-PER-ee-see-FOH-lee-us
hypericifolia,
hypericifolium
With leaves like St John's wort,
Hypericum, as in *Melalaeuca*
hypericifolia

hyssopifolius
hiss-sop-ih-FOH-lee-us
hyssopifolia,
hyssopifolium
With leaves like hyssop,
Hyssopus, as in *Cuphea*
hyssopifolia

ilicifolius
il-liss-ee-FOH-lee-us
ilicifolia, ilicifolium
With leaves like holly, *Ilex*, as in
Itea ilicifolia

juniperifolius
joo-nip-er-ih-FOH-lee-us
juniperifolia,
juniperifolium
With leaves like juniper,
Juniperus, as in *Armeria*
juniperifolia

laurifolius
law-ree-FOH-lee-us
laurifolia, laurifolium
With leaves like bay, *Laurus*, as
in *Cistus laurifolius*

lavandulifolius
lav-an-dew-lih-FOH-lee-us
lavandulifolia,
lavandulifolium
With leaves like lavender,
Lavandula, as in *Salvia*
lavandulifolia

ledifolius
lee-di-FOH-lee-us
ledifolia, ledifolium
With leaves like *Ledum*, as in
Ozothamnus ledifolius

myrtifolius
mir-tih-FOH-lee-us
myrtifolia, myrtifolium
With leaves like *Myrsine*, as in
Leptospermum myrtifolium

napifolius
nap-ih-FOH-lee-us
napifolia, napifolium
With leaves shaped like a turnip,
Brassica rapa (i.e. a flattened
sphere), as in *Salvia napifolia*

pandanifolius
pan-dan-uh-FOH-lee-us
pandanifolia,
pandanifolium
With leaves like *Pandanus*, as in
Eryngium pandanifolium

persicifolius
per-sik-ih-FOH-lee-us
persicifolia, persicifolium
With leaves like the peach tree,
Prunus persica, as in *Campanula*
persicifolia

pinifolius
pin-ih-FOH-lee-us
pinifolia, pinifolium
With leaves like pine, *Pinus*, as
in *Penstemon pinifolius*

porrifolius
po-ree-FOH-lee-us
porrifolia, porrifolium
With leaves like leek, *Allium*
porrum, as in *Tragopogon*
porrifolius

prunifolius
proo-ni-FOH-lee-us
prunifolia, prunifolium
With leaves like cherry, *Prunus*,
as in *Malus prunifolia*

quercifolius
kwer-se-FOH-lee-us
quercifolia, quercifolium
With leaves like oak, *Quercus*, as
in *Hydrangea quercifolia*

The ivy-leaved
speedwell, *Veronica
hederifolia*, is diminutive
in comparison with
common ivy, *Hedera
helix*, but both plants
have a similar leaf shape.

📍 **Though not fruit in the true sense, the seeds of yew, *Taxus baccata*, and the plum-fruited yew, *Prumnopitys taxifolia*, are surrounded by a fleshy covering. In addition, they have similar, two-ranked foliage.**

rosmarinifolius
rose-ma-rih-nih-FOH-lee-us
rosmarinifolia, rosmarinifolium
With leaves like rosemary, *Rosmarinus*, as in *Santolina rosmarinifolia*

ruscifolius
rus-kih-FOH-lee-us
ruscifolia, ruscifolium
With leaves like butcher's broom, *Ruscus*, as in *Sarcococca ruscifolia*

salviifolius
sal-vee-FOH-lee-us
salviifolia, salviifolium
With leaves like sage, *Salvia*, as in *Cistus salviifolius*

sambucifolius
sam-boo-kih-FOH-lee-us
sambucifolia, sambucifolium
With leaves like elder, *Sambucus*, as in *Rodgersia sambucifolia*

serpyllifolius
ser-pil-ly-FOH-lee-us
serpyllifolia, serpyllifolium
With leaves like wild or creeping thyme, *Thymus serpyllum*, as in *Arenaria serpyllifolia*

sonchifolius
son-chi-FOH-lee-us
sonchifolia, sonchifolium
With leaves like sow thistle, *Sonchus*, as in *Francoa sonchifolia*

sorbifolius
sor-bih-FOH-lee-us
sorbifolia, sorbifolium
With leaves like mountain ash, *Sorbus*, as in *Xanthoceras sorbifolium*

tanacetifolius
tan-uh-kee-tih-FOH-lee-us
tanacetifolia, tanacetifolium
With leaves like tansy, *Tanacetum*, as in *Phacelia tanacetifolia*

taxifolius
taks-ih-FOH-lee-us
taxifolia, taxifolium
With leaves like yew, *Taxus*, as in *Prumnopitys taxifolia*

terebinthifolius
ter-ee-binth-ih-FOH-lee-us
terebinthifolia, terebinthifolium
With leaves that smell of turpentine (from the turpentine tree, *Pistacia terebinthus*), as in *Schinus terebinthifolius*

thymifolius
ty-mih-FOH-lee-us
thymifolia, thymifolium
With leaves like thyme, *Thymus*, as in *Lythrum thymifolium*

urticifolius
ur-tik-ih-FOH-lee-us
urticifolia, urticifolium
With leaves like nettle, *Urtica*, as in *Agastache urticifolia*

verbascifolius
ver-bask-ih-FOH-lee-us
verbascifolia, verbascifolium
With leaves like mullein, *Verbascum*, as in *Celmisia verbascifolia*

vitifolius
vy-tih-FOH-lee-us
vitifolia, vitifolium
With leaves like vine, *Vitis*, as in *Abutilon vitifolium*

📍 **The attractive but little-grown yellowhorn, *Xanthoceras sorbifolium*, is a Chinese species of shrub related to maples, *Acer*, and horse chestnuts, *Aesculus*, but its leaves are more reminiscent of those of rowan trees, *Sorbus*.**

CHAPTER FIVE

Places and People

It may not be clear how plant names commemorating people or identifying a place of origin benefit gardeners. However, understanding that *magellanicus* refers to the Strait of Magellan, where the climate is cool and wet, and that *maderensis* indicates the subtropical island of Madeira gives us clues as to the cultivation requirements of plants bearing these names. Similarly, knowing that George Forrest collected plants mainly in temperate China tells us something about how to cultivate plants named *forrestii*.

Europe

The climate of the Pacific Northwest is most suitable for cultivating the flora of Europe, though many European natives thrive in various parts of the US. Much of the flora of Europe can be grown very satisfactorily in cool temperate climates, even those originating from more southerly latitudes as referred to by names such as *lusitanicus* (Portugal and parts of Spain) and *aetnensis* (Mount Etna, Italy).

aetnensis
eet-NEN-sis
aetnensis, aetnense
From Mount Etna, Italy, as in
Genista aetnensis

altaclerensis
al-ta-cler-EN-sis
altaclerensis, altaclerense
From Highclere Castle,
Hampshire, England, as in *Ilex*
× *altaclerensis*

austriacus
oss-tree-AH-kus
austriaca, austriacum
Connected with Austria, as in
Doronicum austriacum

avellanus
av-el-AH-nus
avellana, avellanum
Connected with Avella, Italy, as
in *Corylus avellana*

azoricus
a-ZOR-ih-kus
azorica, azoricum
Connected with the Azores
Islands, as in *Jasminum azoricum*

bannaticus
ban-AT-ih-kus
bannatica, bannaticum
Connected with Banat, Central
Europe, as in *Echinops
bannaticus*

beesianus
bee-zee-AH-nus
beesiana, beesianum
Named after Bees Nursery,
Chester, England, as in *Allium
beesianum*

berolinensis
ber-oh-lin-EN-sis
berolinensis, berolinense
From Berlin, Germany, as in
Populus × *berolinensis*

bodnantense
bod-nan-TEN-see
Named after Bodnant Gardens,
Wales, as in *Viburnum* ×
bodnantense

byzantinus
biz-an-TEE-nus
byzantina, byzantinum
Connected with Istanbul,
Turkey, as in *Colchicum
byzantinum*

cambricus
KAM-brih-kus
cambrica, cambricum
Connected with Wales, as in
Meconopsis cambrica

◄ In the cold, late winter and
early spring the resilient, pink
flowers of round-leaved
cyclamen, *Cyclamen coum*, are a
cheering sight but its name
refers to its warmer origins on
the island of Kos.

cantabricus
kan-TAB-rih-kus
cantabrica, cantabricum
Connected with the Cantabria
region of Spain, as in *Narcissus
cantabricus*

carpaticus
kar-PAT-ih-kus
carpatica, carpaticum
Connected with the Carpathian
Mountains, as in *Campanula
carpatica*

carthusianorum
kar-thoo-see-an-OR-um
Of Grande Chartreuse,
Carthusian monastery near
Grenoble, France, as in *Dianthus
carthusianorum*

chalcedonicus
kalk-ee-DON-ih-kus
chalcedonica,
chalcedonicum
Connected with Chalcedon, the
ancient name for a district of
Istanbul, Turkey, as in *Lychnis
chalcedonica*

clandonensis
klan-don-EN-sis
From Clandon, England, as in
Caryopteris × clandonensis

coum
KOO-um
Connected with Kos, Greece, as
in *Cyclamen coum*

creticus
KRET-ih-kus
cretica, creticum
Connected with Crete, Greece,
as in *Pteris cretica*

◀ The spindle,
Euonymus europaeus,
is a European native
noted for its highly
colorful fruit, which is
pink on the outside
opening to reveal the
orange-coated seeds.

darleyensis
dar-lee-EN-sis
Of Darley Dale nursery (James
Smith & Sons), Derbyshire,
England, as in *Erica × darleyensis*

europaeus
yoo-ROH-pay-us
europaea, europaeum
Connected with Europe, as in
Euonymus europaeus

florentinus
flor-en-TEE-nus
florentina, florentinum
Connected with Florence, Italy,
as in *Malus florentina*

gallicus
GAL-ih-kus
gallica, gallicum
Connected with France, as in
Rosa gallica

garganicus
gar-GAN-ih-kus
garganica, garganicum
Connected with Monte
Gargano, Italy, as in *Campanula
garganica*

graecus
GRAY-kus
graeca, graecum
Of Greece, as in *Fritillaria graeca*

helveticus
hel-VET-ih-kus
helvetica, helveticum
Connected with Switzerland, as
in *Erysimum helveticum*

hibernicus
hy-BER-nih-kus
hibernica, hibernicum
Connected with Ireland, as in
Hedera hibernica

olympicus
oh-LIM-pih-kus
olympica, olympicum
Connected with Mount
Olympus, Greece, as in
Hypericum olympicum

pyrenaeus
py-ren-AY-us
pyrenaea, pyrenaeum

pyrenaicus
py-ren-AY-ih-kus
pyrenaica, pyrenaicum
Connected with the Pyrenees, as
in *Fritillaria pyrenaica*

sarniensis
sarn-ee-EN-sis
sarniensis, sarniense
From the island of Sarnia
(Guernsey), as in *Nerine
sarniensis*

siculus
SIK-yoo-lus
sicula, siculum
From Sicily, Italy, as in
Nectaroscordum siculum

uplandicus
up-LAN-ih-kus
uplandica, uplandicum
Connected with Uppland,
Sweden, as in *Symphytum
× uplandicum*

valentinus
val-en-TEE-nus
valentina, valentinum
Connected with Valencia, Spain,
as in *Coronilla valentina*

vitis-idaea
VY-tiss id-uh-EE-uh
Vine of Mount Ida, as in
Vaccinium vitis-idaea

▪ Sarnia is the ancient
Roman name for
Guernsey, and the
name of the Guernsey
lily, *Nerine sarniensis*,
refers to its abundance
in gardens on Guernsey
island. However,
it is a native plant
of South Africa.

idaeus
eye-DAY-ee-us
idaea, idaeum
Connected with Mount Ida,
as in *Rubus idaeus*

illyricus
il-LEER-ih-kus
illyrica, illyricum
Connected with Illyria, the
name for an area of the western
Balkan peninsula in antiquity, as
in *Gladiolus illyricus*

lusitanicus
loo-si-TAN-ih-kus
lusitanica, lusitanicum
Connected with Lusitania
(Portugal and some parts of
Spain), as in *Prunus lusitanica*

lutetianus
loo-tee-shee-AH-nus
lutetiana, lutetianum
Connected with Lutetia (Paris),
France, as in *Circaea lutetiana*

maderensis
ma-der-EN-sis
maderensis, maderense
From Madeira, as in *Geranium
maderense*

monspessulanus
monz-pess-yoo-LAH-nus
monspessulana,
monspessulanum
Connected with Montpellier,
France, as in *Acer
monspessulanum*

olbius
OL-bee-us
olbia, olbium
Connected with the Îles
d'Hyères (Olbia in Latin),
France, as in *Lavatera olbia*

Rubus idaeus

Vaccinium vitis-idaea

BEHIND THE NAME

The wrong idaea

In naming the raspberry, *Rubus idaeus*, and lingonberry, *Vaccinium vitis-idaea*, the botanist Carl Linnaeus made reference to Mount Ida. However, it is not the well-known Mount Ida of Crete but Mount Ida in northwest Turkey that is referred to. This mountain was studied by the early Greek botanist Theophrastus. However, Linnaeus' use of the name *Vaccinium vitis-idaea* is probably an error for *V. myrtillus* as lingonberry is absent from the mountain. Likewise, the name *Rubus idaeus* is founded on reports of its abundance on Mount Ida made by another ancient Greek botanist, Dioscorides, though Theophrastus did not record it for the area.

105

Asia

Asia covers a huge area from eastern Turkey to the Russian islands in the east. This range incorporates a number of climatic zones but contains a great diversity of plants that make admirable garden specimens in temperate climates. Distance is not necessarily a good index of compatibility, and plants from far away China and Japan (often bearing the names *chinensis* or *japonica*) are mainstays of gardens in many parts of the US.

arabicus
a-RAB-ih-kus
arabica, arabicum
Connected with Arabia, as in
Coffea arabica

armeniacus
ar-men-ee-AH-kus
armeniaca, armeniacum
Connected with Armenia, as in
Muscari armeniacum

babylonicus
bab-il-LON-ih-kus
babylonica, babylonicum
Connected with Babylonia,
Mesopotamia (Iraq), as in *Salix babylonica*, which Linnaeus mistakenly believed to be from the southwest Asia

❧ Coffee, *Coffea arabica*, is native to tropical northeast Africa but was probably first cultivated in nearby Yemen on the Arabian peninsula. It is now grown as a crop in many warmer parts of the world.

baldschuanicus
bald-SHWAN-ih-kus
baldschuanica,
baldschuanicum
Connected with Baljuan,
Turkistan, as in *Fallopia baldschuanica*

cachemiricus
kash-MI-rih-kus
cachemirica,
cachemiricum
Connected with Kashmir, as in
Gentiana cachemirica

camtschatcensis
kam-shat-KEN-sis
camtschatcensis,
camtschatcense

camtschaticus
kam-SHAY-tih-kus
camtschatica,
camtschaticum
From or of the Kamchatka peninsula, Russia, as in
Lysichiton camtschatcensis

cappadocicus
kap-puh-doh-SIH-kus
cappadocica,
cappadocicum
Connected with the ancient province of Cappadocia, Asia Minor (Turkey), as in
Omphalodes cappadocica

cashmerianus
kash-meer-ee-AH-nus
cashmeriana,
cashmerianum

cashmirianus
kash-meer-ee-AH-nus
cashmiriana,
cashmirianum

cashmiriensis
kash-meer-ee-EN-sis
cashmiriensis,
cashmiriense
From or of Kashmir, as in
Cupressus cashmeriana

cathayanus
kat-ay-YAH-nus
cathayana, cathayanum
cathayensis
kat-ay-YEN-sis
cathayensis, cathayense
From or of China, as in
Cardiocrinum cathayana

chinensis
CHI-nen-sis
chinensis, chinense
From China, as in *Stachyurus*
chinensis

cilicicus
kil-LEE-kih-kus
cilicica, cilicicum
Connected with Cilicia (Lesser
Armenia), as in *Colchicum*
cilicicum

colchicus
KOHL-chih-kus
colchica, colchicum
Connected with the coastal
region of the Black Sea, Georgia,
as in *Hedera colchica*

coreanus
kor-ee-AH-nus
coreana, coreanum
Connected with Korea, as in
Hemerocallis coreana

damascenus
dam-ASK-ee-nus
damascena, damascenum
Connected with Damascus,
Syria, as in *Nigella damascena*

formosanus
for-MOH-sa-nus
formosana, formosanum
Connected with Formosa
(Taiwan), as in *Pleione*
formosana

himalayensis
him-uh-lay-EN-is
himalayensis, himalayense
From the Himalaya, as in
Geranium himalayense

hupehensis
hew-pay-EN-sis
hupehensis, hupehense
From Hupeh (Hubei), China, as
in *Sorbus hupehensis*

indicus
IN-dih-kus
indica, indicum
Connected with India; may also
apply to plants originating from
the East Indies or China, as in
Lagerstroemia indica

**❦ The Damask rose, *Rosa* ×
damascena, is a hybrid from
Central Asia. Its name probably
relates to its introduction to
England, which may have been
from Damascus, Syria.**

japonicus
juh-PON-ih-kus
japonica, japonicum
Connected with Japan, as in
Cryptomeria japonica

kamtschaticus
kam-SHAY-tih-kus
kamtschatica,
kamtschaticum
Connected with Kamchatka,
Russia, as in *Sedum*
kamtschaticum

karataviensis
kar-uh-taw-vee-EN-sis
karataviensis, karataviense
From the Karatau mountains,
Kazakhstan, as in *Allium*
karataviense

China is the source of a great number of garden plants grown in temperate parts of the world. The Chinese wisteria, *Wisteria sinensis*, is just one beautiful, if aggressive, example.

koreanus
kor-ee-AH-nus
koreana, koreanum
Connected with Korea, as in
Abies koreana

mongolicus
mon-GOL-ih-kus
mongolica, mongolicum
Connected with Mongolia, as in
Quercus mongolica

nepalensis
nep-al-EN-sis
nepalensis, nepalense
nepaulensis
nep-awl-EN-sis
nepaulensis, nepaulense
From Nepal, as in *Hedera
nepalensis*

nipponicus
nip-PON-ih-kus
nipponica, nipponicum
Connected with Nippon
(Japan), as in *Phyllodoce
nipponica*

persicus
PER-sih-kus
persica, persicum
Connected with Persia (Iran), as
in *Parrotia persica*

ponticus
PON-tih-kus
pontica, ponticum
Connected with Pontus, Asia
Minor (Turkey), as in *Daphne
pontica*

rosa-sinensis
RO-sa sy-NEN-sis
The rose of China, as in *Hibiscus
rosa-sinensis*

sachalinensis
saw-kaw-lin-YEN-sis
sachalinensis, sachalinense
From the island Sakhalin, off the
coast of Russia, as in *Abies
sachalinensis*

sardensis
saw-DEN-sis
sardensis, sardense
Of Sardis (Sart), Turkey, as in
Chionodoxa sardensis

sibiricus
sy-BEER-ih-kus
sibirica, sibiricum
Connected with Siberia, as in
Iris sibirica

sikkimensis
sik-im-EN-sis
sikkimensis, sikkimense
From Sikkim, India, as in
Euphorbia sikkimensis

sinensis
sy-NEN-sis
sinensis, sinense
From China, as in *Corylopsis
sinensis*

syriacus
seer-ee-AH-kus
syriaca, syriacum
Connected with Syria, as in
Asclepias syriaca

szechuanicus

se-CHWAN-ih-kus
szechuanica,
szechuanicum
Connected with Szechuan,
China, as in *Populus szechuanica*

takesimanus

tak-ess-ih-MAH-nus
takesimana, takesimanum
Connected with the Liancourt
Rocks (Takeshima in Japanese),
as in *Campanula takesimana*

taliensis

tal-ee-EN-sis
taliensis, taliense
From the Tali range, Yunnan,
China, as in *Lobelia taliensis*

tanguticus

tan-GOO-tih-kus
tangutica, tanguticum
Connected with the Tangut
region of Tibet, as in *Daphne
tangutica*

tatsienensis

tat-see-en-EN-sis
tatsienensis, tatsienense
From Tatsienlu, China, as in
Delphinium tatsienense

tauricus

TAW-ih-kus
taurica, tauricum
Connected with Taurica
(Crimea), as in *Onosma taurica*

thibetanus

ti-bet-AH-nus
thibetana, thibetanum

thibeticus

ti-BET-ih-kus
thibetica, thibeticum
Connected with Tibet, as in
Rubus thibetanus

yakushimanus

ya-koo-shim-MAH-nus
yakushimana,
yakushimanum
Connected with Yakushima
Island, Japan, as in
Rhododendron yakushimanum

yunnanensis

yoo-nan-EN-sis
yunnanensis, yunnanense
From Yunnan, China, as in
Magnolia yunnanensis

The origins of the rose of
China, *Hibiscus rosa-sinensis*,
are uncertain. It is not now
found in the wild but it is
thought likely to have been
found originally in tropical
Asia, as suggested by its name.

North and South America

North and South America have not contributed as many garden plants as has Asia, though names such as *Picea sitchensis*, the Sitka spruce, from Alaska, and the New York aster, *Symphyotrichum novi-belgii*, will be familiar to many. Some names can be misleading, though. The Portuguese squill, *Scilla peruviana*, hails not from Peru but the Mediterranean, its name referring to a ship called *Peru* on which bulbs were originally imported.

amazonicus
am-uh-ZOH-nih-kus
amazonica, amazonicum
Connected with the Amazon
River, South America, as in
Victoria amazonica

americanus
a-mer-ih-KAH-nus
americana, americanum
Connected with North or South
America, as in *Lysichiton
americanus*

andicola
an-DIH-koh-luh
andinus
an-DEE nus
andina, andinum
Connected with the Andes,
South America, as in *Calceolaria
andina*

araucana
air-ah-KAY-nuh
Relating to the Arauco region in
Chile, as in *Araucaria araucana*

bonariensis
bon-ar-ee-EN-sis
bonariensis, bonariense
From Buenos Aires, as in
Verbena bonariensis

californicus
kal-ih-FOR-nih-kus
californica, californicum
Connected with California, US,
as in *Zauschneria californica*

canadensis
ka-na-DEN-sis
canadensis, canadense
From Canada, though once also
applied to northeastern parts of
the US, as in *Cornus canadensis*

catawbiensis
ka-taw-bee-EN-sis
catawbiensis, catawbiense
From the Catawba River, North
Carolina, US, as in
Rhododendron catawbiense

chilensis
chil-ee-EN-sis
chilensis, chilense
From Chile, as in *Blechnum
chilense*

◄ The giant waterlily, *Victoria
amazonica*, is an extraordinary
plant with lily pads up to 10ft
across. However, its Amazonian
origins mean it can be
cultivated only under glass in
temperate regions.

chiloensis
kye-loh-EN-sis
chiloensis, chiloense
From the island of Chiloé, Chile,
as in *Fragaria chiloensis*

fluminensis
floo-min-EN-sis
fluminensis, fluminense
From Rio de Janeiro, Brazil, as in
Tradescantia fluminensis

jalapa
juh-LAP-a
Connected with Xalapa,
Mexico, as in *Mirabilis jalapa*

ludovicianus
loo-doh-vik-ee-AH-nus
ludoviciana, ludovicianum
Connected with Louisiana, US,
as in *Artemisia ludoviciana*

magellanicus
ma-jell-AN-ih-kus
megallanica,
megallanicum
Connected with the Strait of
Magellan, South America, as in
Fuchsia magellanica

novae-angliae
NO-vee ANG-lee-a
Connected with New England,
US, as in *Symphyotrichum
novae-angliae*

novi-belgii
NO-vee BEL-jee-eye
Connected with New York, US,
as in *Symphyotrichum novi-belgii*

sitchensis
sit-KEN-sis
sitchensis, sitchense
From Sitka, Alaska, US, as in
Sorbus sitchensis

tequilana
te-kee-lee-AH-nuh
Connected with Tequila,
Mexico, as in *Agave tequilana*

tuolumnensis
too-ah-lum-NEN-sis
tuolumnensis,
tuolumnense
From Tuolumne County,
California, US, as in
Erythronium tuolumnense

valdivianus
val-div-ee-AH-nus
valdiviana, valdivianum
Connected with Valdivia, Chile,
as in *Ribes valdivianum*

❦ The four o'clock flower,
Mirabilis jalapa. The purgative
drug jalap derives its name from
the Mexican town of Xalapa
where, historically, it was
commonly prepared. It is a
product of the roots of *Ipomoea
purga*, but Carl Linnaeus falsely
believed it was made from a
species of *Mirabilis* to which he
gave the epithet *jalapa*.
However, it is thought this
widespread and much-
cultivated plant probably does
originate from the Americas.

Other areas of the world

Names relating to Africa or Oceania are surprisingly few, and some of those that appear to can be misleading. The name *australis* means southern rather than referring to Australia, and the commonly grown cabbage tree, *Cordyline australis*, is native to New Zealand. It may be expected that the Tasmanian tree fern, *Dicksonia antarctica*, would display extreme cold-tolerance but, in fact, it is for milder areas only.

abyssinicus
a-biss-IN-ih-kus
abyssinica, abyssinicum
Connected with Abyssinia (Ethiopia), as in *Aponogeton abyssinicus*

aethiopicus
ee-thee-OH-pih-kus
aethiopica, aethiopicum
Connected with Africa, as in *Zantedeschia aethiopica*

africanus
af-ri-KAHN-us
africana, africanum
Connected with Africa, as in *Sparrmannia africana*

algeriensis
al-jir-ee-EN-sis
algeriensis, algeriense
From Algeria, as in *Ornithogalum algeriense*

antarcticus
ant-ARK-tih-kus
antarctica, antarcticum
Connected with the Antarctic region, as in *Dicksonia antarctica*

antipodus
an-te-PO-dus
antipoda, antipodum

antipodeum
an-te-PO-dee-um
Connected with the Antipodes, as in *Gaultheria antipoda*

atlanticus
at-LAN-tih-kus
atlantica, atlanticum
Connected with the Atlantic shoreline, or from the Atlas Mountains, as in *Cedrus atlantica*

australis
aw-STRAH-lis
australis, australe
Southern, as in *Cordyline australis*

borealis
bor-ee-AH-lis
borealis, boreale
Northern, as in *Erigeron borealis*

The red-hot poker tree, *Erythrina abyssinica*, is native to Abyssinia (Ethiopia), as its name suggests, but is also found in other parts of Africa. Its seeds are used to make jewellery but when crushed are poisonous.

Native to damp habitats in South Africa, the calla lily, *Zantedeschia aethiopica*, can be grown outside in mild parts of the US. However, it tends to thrive in somewhat drier soils.

capensis
ka-PEN-sis
capensis, capense
From the Cape of Good Hope, South Africa, as in *Phygelius capensis*

niloticus
nil-OH-tih-kus
nilotica, niloticum
Connected with the Nile Valley, as in *Salvia nilotica*

novae-zelandiae
NO-vay zee-LAN-dee-ay
Connected with New Zealand, as in *Acaena novae-zelandiae*

occidentalis
ok-sih-den-TAH-lis
occidentalis, occidentale
Relating to the West, as in *Thuja occidentalis*

orientalis
or-ee-en-TAH-lis
orientalis, orientale
Relating to the Orient; the East, as in *Thuja orientalis*

The Cape honeysuckle, *Tecoma capensis*, is a scrambling plant native to southern Africa. It is grown in frost-free areas of the world as a hedging plant but should be grown as an annual in cooler parts of the US.

septentrionalis
sep-ten-tree-oh-NAH-lis
septentrionalis, septentrionale
From the north, as in *Beschorneria septentrionalis*

tasmanicus
tas-MAN-ih-kus
tasmanica, tasmanicum
Connected with Tasmania, Australia, as in *Dianella tasmanica*

tingitanus
ting-ee-TAH-nus
tingitana, tingitanum
Connected with Tangiers, as in *Lathyrus tingitanus*

Plant habitats

Successful gardening relies on matching plants with their requirements. There is little point in growing a marginal aquatic in a gravel garden or a desert plant in woodland. Knowing something of a plant's wild habitat, therefore, is essential. A number of names tell us something of plants' needs and tolerances. Species bearing the name *maritimus*, for example, are likely to be tolerant of salt-laden winds but not shade.

agrestis
ag-RES-tis
agrestis, agreste
Found growing in fields, as in *Fritillaria agrestis*

alpestris
al-PES-tris
alpestris, alpestre
Of lower, usually wooded, mountain habitats, as in *Narcissus alpestris*

alpicola
al-PIH-koh-luh
Of high mountain habitats, as in *Primula alpicola*

alpinus
al-PEE-nus
alpina, alpinum
Of high, often rocky regions; from the Alps region of Europe, as in *Pulsatilla alpina*

ammophilus
am-oh-FIL-us
ammophila, ammophilum
Of sandy places, as in *Oenothera ammophila*

amphibius
am-FIB-ee-us
amphibia, amphibium
Growing both on land and in water, as in *Persicaria amphibia*

aquaticus
a-KWA-tih-kus
aquatica, aquaticum

aquatalis
ak-wa-TIL-is
aquatalis, aquatale
Growing in or near water, as in *Mentha aquatica*

arboricola
ar-bor-IH-koh-luh
Living on trees, as in *Schefflera arboricola*

arenarius
ar-en-AH-ree-us
arenaria, arenarium

arenicola
ar-en-IH-koh-luh

arenosus
ar-en-OH-sus
arenosa, arenosum
Growing in sandy places, as in *Leymus arenarius*

The swamp smartweed, *Persicaria amphibia*, is a native plant in Britain and Ireland found in water and other wet places, but it is also a weed on drier, rough ground.

arvensis
ar-VEN-sis
arvensis, arvense
Growing in cultivated fields, as in *Rosa arvensis*

bulbosus
bul-BOH-sus
bulbosa, bulbosum
Bulbous, swollen stem growing underground; resembling a bulb, as in *Ranunculus bulbosus*

campestris
kam-PES-tris
campestris, campestre
Of fields or open plains, as in *Acer campestre*

collinus
kol-EE-nus
collina, collinum
Relating to hills, as in *Geranium collinum*

demersus
DEM-er-sus
demersa, demersum
Living under water, as in *Ceratophyllum demersum*

dendrophilus
den-dro-FIL-us
dendrophila, dendrophilum
Tree-loving, as in *Tecomanthe dendrophila*

deserti
DES-er-tee
Connected with the desert, as in *Agave deserti*

dumetorum
doo-met-OR-um
From hedges or bushes, as in *Fallopia dumetorum*

fluitans
FLOO-ih-tanz
Floating, as in *Glyceria fluitans*

fluvialis
floo-vee-AHL-is
fluvialis, fluviale

fluviatilis
floo-vee-uh-TIL-is
fluviatilis, fluviatile
Growing in a river or running water, as in *Isotoma fluviatilis*

fontanus
FON-tah-nus
fontana, fontanum
Growing in fast-running water, as in *Cerastium fontanum*

frigidus
FRIH-jih-dus
frigida, frigidum
Growing in cold regions, as in *Artemisia frigida*

The field maple, *Acer campestre*, is a native European tree but also cultivated in the US. It is characteristic of scrub and hedgerows defining field boundaries. In gardens it can be similarly used in wildlife hedges.

glacialis
glass-ee-AH-lis
glacialis, glaciale
Connected with ice-cold, glacial regions, as in *Dianthus glacialis*

graniticus
gran-NY-tih-kus
granitica, graniticum
Growing on granite and rocks, as in *Dianthus graniticus*

hypopitys
hi-po-PY-tees
Growing under pines, as in *Monotropa hypopitys*

lacustris
lah-KUS-tris
lacustris, lacustre
Relating to lakes, as in *Iris lacustris*

littoralis
lit-tor-AH-lis
littoralis, littorale

littoreus
lit-TOR-ee-us
littorea, littoreum
Growing by the sea, as in *Griselinia littoralis*

● The rocky, mountain habitats of the west coast of the US that are the favored habitat of the beardtongue, *Penstemon rupicola*, give the plant its name.

maritimus
muh-RIT-tim-mus
maritima, maritimum
Relating to the sea, as in *Armeria maritima*

megapotamicus
meg-uh-poh-TAM-ih-kus
megapotamica, megapotamicum
Connected with a big river: for example the Amazon or Rio Grande, as in *Abutilon megapotamicum*

montanus
MON-tah-nus
montana, montanum
Relating to mountains, as in *Clematis montana*

montensis
mont-EN-sis
montensis, montense

monticola
mon-TIH-koh-luh
Growing on mountains, as in *Halesia monticola*

muralis
mur-AH-lis
muralis, murale
Growing on walls, as in *Cymbalaria muralis*

nemoralis
nem-or-RAH-lis
nemoralis, nemorale

nemorosus
nem-or-OH-sus
nemorosa, nemorosum
Of woodland, as in *Anemone nemorosa*

nubicola
noo-BIH-koh-luh
Growing up in the clouds, as in *Salvia nubicola*

oreophilus
or-ee-O-fil-us
oreophila, oreophilum
Mountain-loving, as in *Sarracenia oreophila*

paludosus
pal-oo-DOH-sus
paludosa, paludosum

palustris
pal-US-tris
palustris, palustre
Of marshland, as in *Quercus palustris*

petraeus
pet-RAY-us
petraea, petraeum
Connected with rocky regions, as in *Quercus petraea*

pinetorum
py-net-OR-um
Connected with pine forests, as in *Fritillaria pinetorum*

pratensis
pray-TEN-sis
pratensis, pratense
From the meadow, as in *Geranium pratense*

riparius
rip-AH-ree-us
riparia, riparium
Of riverbanks, as in *Ageratina riparia*

rivalis
riv-AH-lis
rivalis, rivale
Growing by the side of streams, as in *Geum rivale*

rivularis
riv-yoo-LAH-ris
rivularis, rivulare
Brook-loving, as in *Cirsium rivulare*

rudis
ROO-dis
rudis, rude
Coarse, growing on uncultivated ground, as in *Persicaria rudis*

rupestris
rue-PES-tris
rupestris, rupestre
Of rocky places, as in *Leptospermum rupestre*

rupicola
roo-PIH-koh-luh
Growing on cliffs and ledges, as in *Penstemon rupicola*

saxatilis
saks-A-til-is
saxatilis, saxatile
Of rocky places, as in *Aurinia saxatilis*

saxicola
saks-IH-koh-luh
Growing in rocky places, as in *Juniperus saxicola*

saxifraga
saks-ee-FRAH-gah
Rock-breaking, as in *Petrorhagia saxifraga*

saxorum
saks-OR-um
Of the rocks, as in *Streptocarpus saxorum*

saxosus
saks-OH-sus
saxosa, saxosum
Of rocky places, as in *Gentiana saxosa*

scopulorum
sko-puh-LOR-um
Of crags or cliffs, as in *Cirsium scopulorum*

segetalis
seg-UH-ta-lis
segetalis, segetale
segetum
seg-EE-tum
Of cornfields, as in *Euphorbia segetalis*

sepium
SEP-ee-um
Growing along hedgerows, as in *Calystegia sepium*

subalpinus
sub-al-PY-nus
subalpina, subalpinum
Growing at the lower levels of mountain ranges, as in *Viburnum subalpinum*

submersus
sub-MER-sus
submersa, submersum
Submerged, as in *Ceratophyllum submersum*

sylvaticus
sil-VAT-ih-kus
sylvatica, sylvaticum
sylvester
sil-VESS-ter
sylvestris
sil-VESS-tris
sylvestris, sylvestre
sylvicola
sil-VIH-koh-luh
Growing in woodlands, as in *Pinus sylvestris*

terrestris
ter-RES-tris
terrestris, terrestre
From the ground; growing in the ground, as in *Lysimachia terrestris*

uliginosus
ew-li-gi-NOH-sus
uliginosa, uliginosum
From swampy and wet regions, as in *Salvia uliginosa*

umbrosus
um-BROH-sus
umbrosa, umbrosum
Growing in shade, as in *Phlomis umbrosa*

urbanus
ur-BAH-nus
urbana, urbanum
urbicus
UR-bih-kus
urbica, urbicum
urbius
UR-bee-us
urbia, urbium
From cities, as in *Geum urbanum*

◄ In the wild meadow cranesbill, *Geranium pratense*, grows among other herbaceous species, making it a perfect border plant. Some fine selections are available and it is also the parent of a number of hybrids.

People

The names of some plants commemorating people tell us little about them but others can be revealing. For example, Ernest Wilson collected so many garden plants from China that he earned the sobriquet "Chinese," while the eminent botanist Sir Joseph Banks lent his name to a number of species from Oceania. Researching this kind of information transcends the boundaries of the garden.

archangelica
ark-an-JEL-ih-kuh
In reference to the Archangel Raphael, as in *Angelica archangelica*

arendsii
ar-END-see-eye
Named after Georg Arends (1862–1952), German nurseryman, as in *Astilbe × arendsii*

armandii
ar-MOND-ee-eye
Named after Armand David (1826–1900), French naturalist and missionary, as in *Pinus armandii*

banksianus
banks-ee-AH-nus
banksiana, banksianum

banksii
BANK-see-eye
Named after Sir Joseph Banks (1743–1820), English botanist and plant collector, as in *Cordyline banksii*. The epithet *banksiae* commemorates his wife, Lady Dorothea Banks (1758–1828)

berthelotii
berth-eh-LOT-ee-eye
Named after Sabin Berthelot (1794–1880), French naturalist, as in *Lotus berthelotii*

bodinieri
boh-din-ee-ER-ee
Named after Émile-Marie Bodinier (1842–1901), French missionary who collected plants in China, as in *Callicarpa bodinieri*

bowdenii
bow-DEN-ee-eye
Named after plantsman Athelstan Cornish-Bowden (1871–1942), as in *Nerine bowdenii*

bulleyanus
bul-ee-YAH-nus
bulleyana, bulleyanum

bulleyi
bul-ee-YAH-eye
Named after Arthur Bulley (1861–1942), founder of Ness Botanic Gardens, Cheshire, England, as in *Primula bulleyana*

bungeanus
bun-jee-AH-nus
bungeana, bungeanum

bungei
bun-jee-EYE
Named after Dr. Alexander von Bunge (1803–90), Russian botanist, as in *Pinus bungeana*

burkwoodii
berk-WOOD-ee-eye
Named after brothers Arthur and Albert Burkwood, 19th-century hybridizers, as in *Viburnum × burkwoodii*

calleryanus
kal-lee-ree-AH-nus
calleryana, calleryanum
Named after Joseph-Marie Callery (1810–62), French missionary who plant-hunted in France, as in *Pyrus calleryana*

colensoi
co-len-SO-ee
Named after the Revd William Colenso (1811–99), New Zealand plant collector, as in *Pittosporum colensoi*

coulteri
kol-TER-ee-eye
Named after Dr. Thomas Coulter (1793–1843), Irish botanist, as in *Romneya coulteri*

darwinii
dar-WIN-ee-eye
Named after Charles Darwin
(1809–82), English naturalist, as
in *Berberis darwinii*

delavayi
del-uh-VAY-ee
Named after Père Jean Marie
Delavay (1834–95), French
missionary, explorer, and
botanist, as in *Magnolia delavayi*

douglasianus
dug-lus-ee-AH-nus
douglasiana,
douglasianum
douglasii
dug-lus-EE-eye
Named after David Douglas
(1799–1834), Scottish plant
hunter, as in *Limnanthes
douglasii*

elwesii
el-WEZ-ee-eye
Named after Henry John Elwes
(1846–1922), British plant
collector, one of the inaugural
recipients of the Victoria Medal
of the Royal Horticultural
Society, as in *Galanthus elwesii*

endresii
en-DRESS-ee-eye
endressii
Named after Philip Anton
Christoph Endress (1806–31),
German plant collector, as in
Geranium endressii

◄ The Revd William
Colenso was born in
Cornwall but moved to
New Zealand as a young
man and studied the flora
he found there. He is
commemorated in the
names of many New
Zealand plants including
rautawhiri, *Pittosporum
colensoi*.

farreri
far-REY-ree
Named after Reginald Farrer
(1880–1920), English plant
hunter and botanist, as in
Viburnum farreri

forrestianus
for-rest-ee-AH-nus
forrestiana, forrestianum
forrestii
for-rest-EE-eye
Named after George Forrest
(1873–1932), Scottish plant
collector, as in *Hypericum
forrestii*

fortunei
for-TOO-nee-eye
Named after Robert Fortune
(1812–80), Scottish plant hunter
and horticulturist, as in
Trachycarpus fortunei

haastii
HAAS-tee-eye
Named after Sir Julius von Haast
(1824–87), German explorer
and geologist, as in *Olearia* ×
haastii

henryi
HEN-ree-eye
Named after Augustine Henry
(1857–1930), Irish plant
collector, as in *Lilium henryi*

leichtlinii
leekt-LIN-ee-eye
Named after Max Leichtlin
(1831–1910), German plant
collector from Baden-Baden,
Germany, as in *Camassia
leichtlinii*

lindleyanus
lind-lee-AH-nus
lindleyana, lindleyanum
lindleyi
lind-lee-EYE
Named after John Lindley
(1799–1865), English botanist
associated with the Royal
Horticultural Society, as in
Buddleja lindleyana

linnaeanus
lin-ee-AH-nus
linnaeana, linnaeanum
linnaei
lin-ee-EYE
Named after Carl Linnaeus
(1707–78), Swedish botanist, as
in *Solanum linnaeanum*

marianus
mar-ee-AH-nus
mariana, marianum
Of the Virgin Mary (or
sometimes Maryland, US),
as in *Silybum marianum*

menziesii
menz-ESS-ee-eye
Named after Archibald Menzies
(1754–1842), British naval
surgeon and botanist, as in
Pseudotsuga menziesii

pernyi
PERN-yee-eye
Named after Paul Hubert Perny
(1818–1907), French missionary
and botanist, as in *Ilex pernyi*

purpusii
pur-PUSS-ee-eye
Named after Carl Purpus
(1851–1941) or his brother
Joseph Purpus (1860–1932),
German plant collectors, as in
Lonicera × purpusii

rehderi
REH-der-eye
rehderianus
re-der-ee-AH-nus
rehderiana, rehderianum
Named after Alfred Rehder
(1863–1949), German-born
dendrologist who worked at the
Arnold Arboretum,
Massachusetts, US, as in
Clematis rehderiana

◄ Williams' camellia,
Camellia × williamsii, is a
hybrid between a Chinese and a
Japanese species that was raised
at Caerhays Castle in Cornwall,
England, in the 1920s. A huge
range of cultivars is now
available and they are
invaluable garden shrubs.

rockii
ROK-ee-eye
Named after Joseph Francis
Charles Rock (1884–1962),
Austrian-born American plant
hunter, as in *Paeonia rockii*

sargentianus
sar-jen-tee-AH-nus
sargentiana, sargentianum
sargentii
sar-JEN-tee-eye
Named after Charles Sprague
Sargent (1841–1927),
dendrologist and director of the
Arnold Arboretum,
Massachusetts, US, as in *Sorbus
sargentiana*

sieboldianus
see-bold-ee-AH-nus
sieboldiana, sieboldianum
sieboldii
see-bold-ee-EYE
Named after Philipp von Siebold
(1796–1866), German doctor
who collected plants in Japan, as
in *Magnolia sieboldii*

sprengeri
SPRENG-er-ee
Named after Carl Ludwig
Sprenger (1846–1917), German
botanist and plantsman, who
bred and introduced many new
plants, as in *Tulipa sprengeri*

thunbergii
thun-BERG-ee-eye
Named after Carl Peter
Thunberg (1743–1828), Swedish
botanist, as in *Spiraea thunbergii*

tommasinianus
toh-mas-see-nee-AH-nus
tommasiniana,
tommasinianum
Named after Muzio Giuseppe
Spirito de' Tommasini
(1794–1879), Italian botanist, as
in *Campanula tommasiniana*

torreyanus
tor-ree-AH-nus
torreyana, torreyanum
Named after Dr. John Torrey
(1796–1873), American
botanist, as in *Pinus torreyana*

traversii
trav-ERZ-ee-eye
Named after William Travers
(1819–1903), New Zealand
lawyer, politician, and plant
collector, as in *Celmisia traversii*

tschonoskii
chon-OSK-ee-eye
Named after Sugawa Tschonoski
(1841–1925), Japanese botanist
and plant collector, as in *Malus
tschonoskii*

turczaninowii
tur-zan-in-NOV-ee-eye
Named after Nicholai S.
Turczaninov (1796–1863),
Russian botanist, as in *Carpinus
turczaninowii*

urvilleanus
ur-VIL-ah-nus
urvilleana, urvilleanum
Named after J.S.C. Dumont
d'Urville (1790–1842), French
botanist and explorer, as in
Tibouchina urvilleana

veitchianus
veet-chee-AH-nus
veitchiana, veitchianum
veitchii
veet-chee-EYE
Named after members of the
Veitch family, nurserymen of
Exeter and Chelsea, as in *Paeonia
veitchii*

vilmorinianus
vil-mor-in-ee-AH-nus
vilmoriniana,
vilmorinianum
vilmorinii
vil-mor-IN-ee-eye
Named after Maurice de
Vilmorin (1849–1918),
French nurseryman, as in
Cotoneaster vilmorinianus

wallichianus
wal-ik-ee-AH-nus
wallichiana, wallichianum
Named after Dr. Nathaniel
Wallich (1786–1854), Danish
botanist and plant hunter, as in
Pinus wallichiana

williamsii
wil-yams-EE-eye
Named for various eminent
botanists and horticulturists
called Williams, including John
Charles Williams (1861–1939),
English plant collector, as in
Camellia × williamsii

wilsoniae
wil-SON-ee-ay
wilsonii
wil-SON-ee-eye
Named after Dr. Ernest Henry
Wilson (1876–1930), English
plant hunter, as in *Spiraea
wilsonii*. The epithet *wilsoniae*
commemorates his wife Helen

wittrockianus
wit-rok-ee-AH-nus
wittrockiana,
wittrockianum
Named after Professor Veit
Brecher Wittrock (1839–1914),
Swedish botanist, as in *Viola ×
wittrockiana*

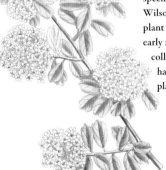

Wilson's bridal wreath,
Spiraea wilsonii, owes its
specific epithet to Ernest
Wilson, one of the great
plant collectors of the
early 20th century, whose
collections in East Asia
have given many beautiful
plants to temperate gardens.

CHAPTER SIX

Ideas, Associations, and Properties

The study of plants is a human endeavour and, as such, the names given to plants are subject to the influence of human constructs, ideas, and values. Names that refer to such concepts make plants relevant to culture as well as nature, and much of interest to gardeners can be learnt from them, from flowering season to physical appearance to properties of practical use. Perhaps most treasured are those names that recognize traditional associations connecting us to our past.

Animal associations

Some plants have strong associations with particular animals. The Venus fly trap, *Dionaea muscipula*, is inextricably linked to its ability to eat flies, a characteristic acknowledged in its name. Similarly, the effect of catmint, *Nepeta cataria*, on cats could hardly be overlooked when naming it. Beware, though, the name of the dog rose, *Rosa canina*. This does not refer to the animal, but to the plant's perceived inferiority to its garden relatives.

aucuparius
awk-yoo-PAH-ree-us
aucuparia, aucuparium
Of bird-catching, as in *Sorbus aucuparia*

avium
AY-ve-um
Relating to birds, as in *Prunus avium*

bufonius
buf-OH-nee-us
bufonia, bufonium
Relating to toads; grows in damp places, as in *Juncus bufonius*

caninus
kay-NEE-nus
canina, caninum
Relating to dogs, often meaning inferior, as in *Rosa canina*

capreus
KAP-ray-us
caprea, capreum
Relating to goats, as in *Salix caprea*

● This highly attractive horse moth orchid, *Phalaenopsis equestris*, makes a good house plant and probably received its name from a supposed similarity between the poise of its flowers and a rider on a horse.

caprifolius
kap-rih-FOH-lee-us
caprifolia, caprifolium
Literally goat leaf, as in *Lonicera caprifolium*

cataria
kat-AR-ee-uh
Relating to cats, as in *Nepeta cataria*

draco
DRAY-koh
Dragon, as in *Dracaena draco*

equestris
e-KWES-tris
equestris, equestre

equinus
e-KWEE-nus
equina, equinum
Relating to horses; equestrian, as in *Phalaenopsis equestris*

muscipula
musk-IP-yoo-luh
Catches flies, as in *Dionaea muscipula*

muscivorus
mus-SEE-ver-us
muscivora, muscivorum
Appearing to eat flies, as in *Helicodiceros muscivorus*

ovinus
oh-VIN-us
ovina, ovinum
Relating to sheep or sheep feed, as in *Festuca ovina*

uva-ursi
OO-va UR-see
Bear's grape, as in *Arctostaphylos uva-ursi*

The Venus fly trap, *Dionaea muscipula*, is a native of North and South Carolina that gains nutrients from insects caught with its highly modified leaves. Despite its specialized lifestyle it adapts readily to cultivation.

A fine, vigorous climber, the perfoliate honeysuckle, *Lonicera caprifolium*, is traditionally held to be a favored food of goats but is also a popular plant in gardens, grown for its fragrant flowers.

125

Useful properties

In the past it has been judged helpful to use plant names to denote potentially useful properties. To this day, many herb garden plants bear the names *officinalis* or *officinarum*, referring to their availability in shops (and therefore their usefulness). Putting into practise the remedies and properties attributed to plants in the past is not recommended, but having knowledge of the lore of garden plants adds to our appreciation of them.

bronchialis
bron-kee-AL-lis
bronchialis, bronchiale
Used in the past as a treatment for bronchitis, as in *Saxifraga bronchialis*

catharticus
kat-AR-tih-kus
carthartica, catharticum
Cathartic; purgative, as in *Rhamnus cathartica*

cerealis
ser-ee-AH-lis
cerealis, cereale
Relating to agriculture; derived from Ceres, the goddess of farming, as in *Secale cereale*

coronarius
kor-oh-NAH-ree-us
coronaria, coronarium
Used for garlands, as in *Anemone coronaria*

elasticus
ee-LASS-tih-kus
elastica, elasticum
Elastic; producing latex, as in *Ficus elastica*

febrifugus
feb-ri-FEW-gus
febrifuga, febrifugum
Can reduce fever, as in *Dichroa febrifuga*

inebrians
in-ee-BRI-enz
Intoxicating, as in *Ribes inebrians*

medicus
MED-ih-kus
medica, medicum
Medicinal, as in *Citrus medica*

officinalis
oh-fiss-ih-NAH-lis
officinalis, officinale
Sold in stores, hence denoting a useful plant (vegetable, culinary or medicinal herb), as in *Rosmarinus officinalis*

officinarum
off-ik-IN-ar-um
From a store, usually an apothecary, as in *Mandragora officinarum*

oleraceus
awl-lur-RAY-see-us
oleracea, oleraceum
Used as a vegetable, as in *Spinacia oleracea*

With a name meaning "used for garlands," the poppy anemone, *Anemone coronaria*, is a showy, tuberous perennial that is often best lifted at the end of the growing season and overwintered in a frost-free place.

papyrifer
pap-IH-riff-er
papyriferus
pap-ih-RIH-fer-us
papyrifera, papyriferum
Producing paper, as in
Tetrapanax papyrifer

papyrus
pa-PY-rus
Ancient Greek word for paper,
as in *Cyperus papyrus*

podagraria
pod-uh-GRAR-ee-uh
From the Latin *podagra*, gout, as
in *Aegopodium podagraria*

sativus
sa-TEE-vus
sativa, sativum
Cultivated, as in *Castanea sativa*

siphiliticus
sigh-fy-LY-tih-kus
siphilitica, siphiliticum
Connected with syphilis, as in
Lobelia siphilitica

somniferus
som-NIH-fer-us
somnifera, somniferum
Inducing sleep, as in *Papaver
somniferum*

textilis
teks-TIL-is
textilis, textile
Relating to weaving, as in
Bambusa textilis

tinctorius
tink-TOR-ee-us
tinctoria, tinctorium
Used as a dye, as in *Genista
tinctoria*

viniferus
vih-NIH-fer-us
vinifera, viniferum
Producing wine, as in *Vitis
vinifera*

vomitorius
vom-ih-TOR-ee-us
vomitoria, vomitorium
Emetic, as in *Ilex vomitoria*

❛ Having been cultivated
since ancient times in southern
Europe, the grape vine, *Vitis
vinifera*, is now found on all
continents except Antarctica. It
grows well in many parts of the
US, but particularly in
California and the Finger Lakes
area of New York state.

Good qualities

It is in the nature of gardeners to invest their plants with personalities and people will often talk of one plant being "happy" or another "sulking." In names, too, human qualities sometimes feature and here are listed those which have positive associations. These are often intuitively understood and as such we know exactly what it means for a tree to be noble (*nobilis*) or a tulip to be distinguished (*praestans*).

amabilis
am-AH-bih-lis
amabilis, amabile
Lovely, as in *Cynoglossum amabile*

amoenus
am-oh-EN-us
amoena, amoenum
Pleasant; delightful, as in *Lilium amoenum*

basilicus
bass-IL-ih-kus
basilica, basilicum
With princely or royal properties, as in *Ocimum basilicum*

blandus
BLAN-dus
blanda, blandum
Mild or charming, as in *Anemone blanda*

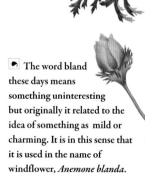

📌 The word bland these days means something uninteresting but originally it related to the idea of something as mild or charming. It is in this sense that it is used in the name of windflower, *Anemone blanda*.

conspicuus
kon-SPIK-yoo-us
conspicua, conspicuum
Conspicuous, as in *Sinningia conspicua*

decoratus
dek-kor-RAH-tus
decorata, decoratum
decorus
dek-kor-RUS
decora, decorum
Decorative, as in *Rhododendron decorum*

elegans
el-ee-GANS
elegantulus
el-eh-GAN-tyoo-lus
elegantula, elegantulum
Elegant, as in *Desmodium elegans*

elegantissimus
el-ee-gan-TISS-ih-mus
elegantissima, elegantissimum
Very elegant, as in *Schefflera elegantissima*

eminens
EM-in-enz
Eminent; prominent, as in *Sorbus eminens*

excellens
ek-SEL-lenz
Excellent, as in *Sarracenia × excellens*

facetus
fa-CEE-tus
faceta, facetum
Elegant, as *Rhododendron facetum*

fecundus
feh-KUN-dus
fecunda, fecundum
Fertile; fruitful, as in *Aeschynanthus fecundus*

fertilis
fer-TIL-is
fertilis, fertile
With plenty of fruit; with many seeds, as in *Robinia fertilis*

festalis
FES-tuh-lis
festalis, festale

festivus
fes-TEE-vus
festiva, festivum
Festive; bright, as in *Hymenocallis × festalis*

firmatus
fir-MAH-tus
firmata, firmatum

firmus
fir-MUS
firma, firmum
Strong, as in *Abies firma*

formosus
for-MOH-sus
formosa, formosum
Handsome; beautiful, as in *Pieris formosa*

gloriosus
glo-ree-OH-sus
gloriosa, gloriosum
Superb; glorious, as in *Yucca gloriosa*

◆ The crown imperial, *Fritillaria imperialis*, is a spectacular bulbous species, its brightly colored flowers topped by a crown of leafy bracts. However, its growth is accompanied by a foxy smell that some find unpleasant.

gracilis
GRASS-il-is
gracilis, gracile
Graceful; slender, as in *Geranium gracile*

grandis
gran-DIS
grandis, grande
Big; showy, as in *Licuala grandis*

imperialis
im-peer-ee-AH-lis
imperialis, imperiale
Very fine; showy, as in *Fritillaria imperialis*

incomparabilis
in-kom-par-RAH-bih-lis
incomparabilis, incomparabile
Incomparable, as in *Narcissus × incomparabilis*

jucundus
joo-KUN-dus
jucunda, jucundum
Agreeable; pleasing, as in *Osteospermum jucundum*

laetus
LEE-tus
laeta, laetum
Bright; vivid, as in *Neopanax laetus*

👁 When brushed against, the leaves of the sensitive plant, *Mimosa pudica*, have the charming habit of folding up. This is a defense mechanism against browsing herbivores but gives the impression of modesty or shyness.

laudatus
law-DAH-tus
laudata, laudatum
Worthy of praise, as in *Rubus laudatus*

lucens
LOO-senz

lucidus
LOO-sid-us
lucida, lucidum
Bright; shining; clear, as in *Ligustrum lucidum*

magnificus
mag-NIH-fih-kus
magnifica, magnificum
Splendid; magnificent, as in *Geranium × magnificum*

mirabilis
mir-AH-bih-lis
mirabilis, mirabile
Wonderful; remarkable, as in *Puya mirabilis*

mitis
MIT-is
mitis, mite
Mild; gentle; without spines, as in *Caryota mitis*

modestus
mo-DES-tus
modesta, modestum
Modest, as in *Aglaonema modestum*

nobilis
NO-bil-is
nobilis, nobile
Noble; renowned, as in *Laurus nobilis*

ornans
OR-nanz

ornatus
or-NA-tus
ornata, ornatum
Ornamental; showy, as in *Musa ornata*

praestans
PRAY-stanz
Distinguished, as in *Tulipa praestans*

princeps
PRIN-keps
Most distinguished, as in *Centaurea princeps*

pudicus
pud-IH-kus
pudica, pudicum
Shy, as in *Mimosa pudica*

pulchellus
pul-KELL-us
pulchella, pulchellum

pulcher
PUL-ker
pulchra, pulchrum
Pretty; beautiful, as in *Correa pulchella*

pulcherrimus
pul-KAIR-ih-mus
pulcherrima, pulcherrimum
Very beautiful, as in *Dierama pulcherrimum*

regalis
re-GAH-lis
regalis, regale
Regal; of exceptional merit, as in *Osmunda regalis*

regius
REE-jee-us
regia, regium
Royal, as in *Juglans regia*

rex
reks
King; with outstanding qualities,
as in *Begonia rex*

speciosus
spee-see-OH-sus
speciosa, speciosum
Showy, as in *Ribes speciosum*

spectabilis
speck-TAH-bih-lis
spectabilis, spectabile
Spectacular; showy, as in *Sedum
spectabile*

splendens
SPLEN-denz
splendidus
splen-DEE-dus
splendida, splendidum
Splendid, as in *Fuchsia splendens*

superbiens
soo-PER-bee-enz
superbus
soo-PER-bus
superba, superbum
Superb, as in *Salvia × superba*

venustus
ven-NUSS-tus
venusta, venustum
Handsome, as in *Hosta venusta*

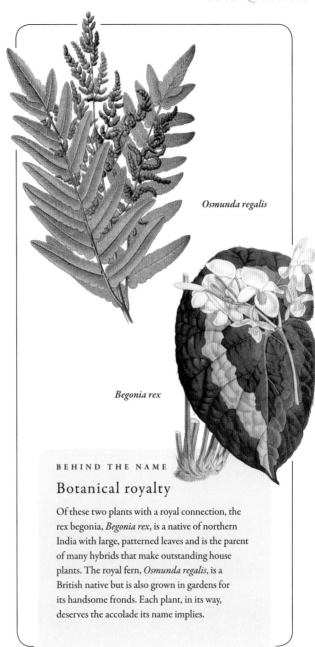

Osmunda regalis

Begonia rex

BEHIND THE NAME
Botanical royalty

Of these two plants with a royal connection, the
rex begonia, *Begonia rex*, is a native of northern
India with large, patterned leaves and is the parent
of many hybrids that make outstanding house
plants. The royal fern, *Osmunda regalis*, is a
British native but is also grown in gardens for
its handsome fronds. Each plant, in its way,
deserves the accolade its name implies.

Bad qualities

Some plant names have negative connotations that can act as a warning. The sap of the upas tree, *Antiaris toxicaria*, has been used to tip poison darts and its name denotes its toxic properties. Equally, there is something dingy about the foliage of brass buttons, *Leptinella squalida*, as its name suggests. But usually it is best to give plants a chance; the night gladiolus, *Gladiolus tristis*, is not as gaudy as some of its relatives, but is very pretty.

controversus
kon-troh-VER-sus
controversa, controversum
Controversial; doubtful, as in *Cornus controversa*

debilis
deb-IL-is
debilis, debile
Weak and frail, as in *Asarum debile*

Ironically, the reason why the great Swedish botanist Carl Linnaeus named the sunset foxglove, *Digitalis obscura*, as he did is somewhat obscure, but he may have been referring to the rather uncertain color of the flowers.

decipiens
de-SIP-ee-enz
Deceptive; not obvious, as in *Sorbus decipiens*

diabolicus
dy-oh-BOL-ih-kus
diabolica, diabolicum
Devilish, as in *Acer diabolicum*

dolosus
do-LOH-sus
dolosa, dolosum
Deceitful; looking like another plant, as in *Cattleya* × *dolosa*

fallax
FAL-laks
Deceptive; false, as in *Crassula fallax*

fastuosus
fast-yoo-OH-sus
fastuosa, fastuosum
Proud, as in *Cassia fastuosa*

fatuus
FAT-yoo-us
fatua, fatuum
Insipid; poor quality, as in *Avena fatua*

futilis
FOO-tih-lis
futilis, futile
Without use, as in *Salsola futilis*

illinitus
il-lin-EYE-tus
illinita, illinitum
Smeared, as in *Escallonia illinita*

infestus
in-FES-tus
infesta, infestum
Dangerous; troublesome, as in *Melilotus infestus*

infortunatus
in-for-tu-NAH-tus
infortunata, infortunatum
Unfortunate (of poisonous plants), as in *Clerodendrum infortunatum*

inquinans
in-KWIN-anz
Polluted; stained; defiled, as in *Pelargonium inquinans*

obscurus
ob-SKEW-rus
obscura, obscurum
Not clear; not certain, as in *Digitalis obscura*

sordidus
SOR-deh-dus
sordida, sordidum
Dirty-looking, as in *Salix × sordida*

spurius
SPEW-eee-us
spuria, spurium
False; spurious, as in *Iris spuria*

squalidus
SKWA-lee-dus
squalida, squalidum
Dirty-looking; dingy, as in *Leptinella squalida*

toxicarius
toks-ih-KAH-ree-us
toxicaria, toxicarium
Poisonous, as in *Antiaris toxicaria*

tristis
TRIS-tis
tristis, triste
Dull; sad, as in *Gladiolus tristis*

urens
UR-enz
Stinging; burning, as in *Urtica urens*

venenosus
ven-ee-NOH-sus
venenosa, venenosum
Very poisonous, as in *Caralluma venenosa*

When naming the false iris, *Iris spuria*, Carl Linnaeus thought it may represent a hybrid rather than a true species. The name reflects this idea that it was false or not as it seemed.

Relationships

The branch of botany dealing with the classification of plants is called taxonomy. In categorizing plants, taxonomists are much-concerned with reflecting relationship. Hybrids between two species are sometimes recognized with names such as *hybridus*, *medius*, *mixtus*, and *intermedius*, and plants bearing these names are often robust and floriferous, making them excellent horticultural subjects.

affinis
uh-FEE-nis
affinis, affine
Related or similar to, as in
Dryopteris affinis

anomalus
ah-NOM-uh-lus
anomala, anomalum
Unlike the norm found in a genus, as in *Hydrangea anomala*

commixtus
kom-MIKS-tus
commixta, commixtum
Mixed; mingled together, as in
Sorbus commixta

commutatus
kom-yoo-TAH-tus
commutata, commutatum
Changed: for example, when formerly included in another species, as in *Papaver commutatum*

confusus
kon-FEW-sus
confusa, confusum
Confused; uncertain, as in
Sarcococca confusa

consanguineus
kon-san-GWIN-ee-us
consanguinea, consanguineum
Related, as in *Vaccinium consanguineum*

decipiens
de-SIP-ee-enz
Deceptive; not obvious, as in
Sorbus decipiens

dubius
DOO-bee-us
dubia, dubium
Doubtful; unlike the rest of the genus, as in *Ornithogalum dubium*

hybridus
hy-BRID-us
hybrida, hybridum
Mixed; hybrid, as in *Helleborus* × *hybridus*

incertus
in-KER-tus
incerta, incertum
Doubtful; uncertain, as in *Draba incerta*

intermedius
in-ter-MEE-dee-us
intermedia, intermedium
Intermediate in color, form or habit, as in *Forsythia* × *intermedia*

medius
MEED-ee-us
media, medium
Intermediate; middle, as in
Mahonia × *media*

mixtus
MIKS-tus
mixta, mixtum
Mixed, as in *Potentilla* × *mixta*

obtectus
ob-TEK-tus
obtecta, obtectum
Covered; protected, as in
Cordyline obtecta

propinquus
prop-IN-kwus
propinqua, propinquum
Related to; near, as in
Myriophyllum propinquum

similis
SIM-il-is
similis, simile
Similar; like, as in *Lonicera similis*

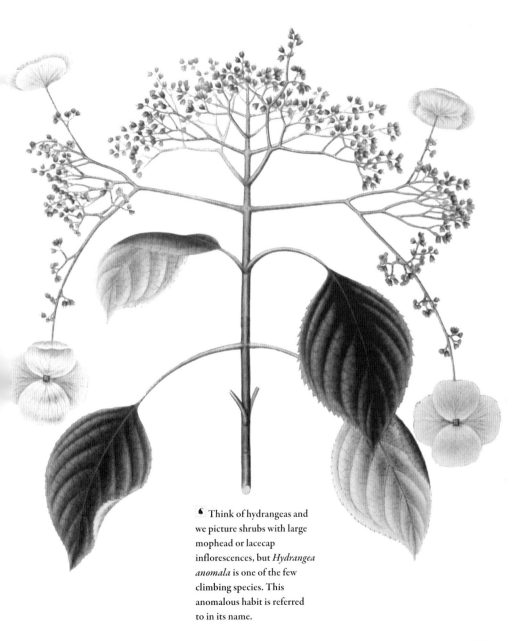

❝ Think of hydrangeas and we picture shrubs with large mophead or lacecap inflorescences, but *Hydrangea anomala* is one of the few climbing species. This anomalous habit is referred to in its name.

Aspects of time

Patience is certainly a virtue for gardeners and observing the progression of the year is one of gardening's chief joys. Most people can call to mind early harbingers of spring such as snowdrops, *Galanthus*, pushing through the frost, or the catkins of hazel, *Corylus*, dancing in the hedgerows, but other plants can be more difficult to place in their season. Fortunately, some plant names can offer a clue.

aestivalis
ee-stiv-AH-lis
aestivalis, aestivale
Relating to summer, as in *Vitis aestivalis*

aestivus
EE-stiv-us
aestiva, aestivum
Developing or ripening in the summer months, as in *Leucojum aestivum*

annuus
AN-yoo-us
annua, annuum
Annual, as in *Helianthus annuus*

autumnalis
aw-tum-NAH-lis
autumnalis, autumnale
Relating to fall (autumn), as in *Colchicum autumnale*

◀ The sunflower, *Helianthus annuus*, is one of the most impressive annual plants, capable of growing several feet in its single year of life. Because many of its relatives are perennials, its life cycle is noted in its name.

biennis
by-EN-is
biennis, bienne
Biennial, as in *Oenothera biennis*

hiemalis
hy-EH-mah-lis
hiemalis, hiemale
Of the winter; winter-flowering, as in *Leucojum hiemale*

hyemalis
hy-EH-mah-lis
hyemalis, hyemale
Relating to winter; winter-flowering, as in *Eranthis hyemalis*

majalis
maj-AH-lis
majalis, majale
Flowering in May, as in *Convallaria majalis*

matronalis
mah-tro-NAH-lis
matronalis, matronale
Relating to March 1st, the Roman Matronalia, festival celebrating Juno as goddess of childbirth and motherhood, as in *Hesperis matronalis*

noctiflorus
nok-tee-FLOR-us
noctiflora, noctiflorum

nocturnus
NOK-ter-nus
nocturna, nocturnum
Flowering at night, as in *Silene noctiflora*

nyctagineus
nyk-ta-JEE-nee-us
nyctaginea, nyctagineum
Flowering at night, as in
Mirabilis nyctaginea

perennis
per-EN-is
perennis, perenne
Perennial, as in *Bellis perennis*

praecox
PRAY-koks
Very early, as in *Stachyurus praecox*

serotinus
se-roh-TEE-nus
serotina, serotinum
With flowers or fruit late in the season, as in *Iris serotina*

tardiflorus
tar-dee-FLOR-us
tardiflora, tardiflorum
Flowering late in the season, as in *Cotoneaster tardiflorus*

tardus
TAR-dus
tarda, tardum
Late, as in *Tulipa tarda*

transitorius
tranz-ee-TAW-ree-us
transitoria, transitorum
Short-lived, as in *Malus transitoria*

trimestris
try-MES-tris
trimestris, trimestre
Of three months, as in *Lavatera trimestris*

veris
VER-is
Relating to spring; flowering in spring, as in *Primula veris*

vernalis
ver-NAH-lis
vernalis, vernale
Relating to spring; flowering in spring, as in *Pulsatilla vernalis*

vernus
VER-nus
verna, vernum
Relating to spring, as in *Leucojum vernum*

❛ The annual mallow, *Lavatera trimestris*, is an exceptionally floriferous plant that is thought to owe its name (meaning "of three months") to the length of its flowering season.

Further concepts

All manner of ideas and allusions appear in plant names, some romantic, some playful, some almost sinister. This section groups together those names that would not sit comfortably in any of the other categories of this book, such as the touch-me-not balsam, *Impatiens noli-tangere*, which received its name, meaning "touch not," for its exploding seed pods, or mourning cypress, *Cupressus funebris*, which is associated with graveyards.

aemulans
EM-yoo-lanz

aemulus
EM-yoo-lus
aemula, aemulum
Imitating; rivalling, as in
Scaevola aemula

aequalis
ee-KWA-lis
aequalis, aequale
Equal, as in *Phygelius aequalis*

anceps
AN-seps
Two-sided; ambiguous, as in
Laelia anceps

barbarus
BAR-bar-rus
barbara, barbarum
Foreign, as in *Lycium barbarum*

ceriferus
ker-IH-fer-us
cerifera, ceriferum
Producing wax, as in *Morella
cerifera*

clandestinus
klan-des-TEE-nus
clandestina, clandestinum
Hidden; concealed, as in
Lathraea clandestina

communis
KOM-yoo-nis
communis, commune
Growing in groups; common, as
in *Myrtus communis*

cultorum
kult-OR-um
Relating to gardens, as in *Trollius
× cultorum*

deciduus
dee-SID-yu-us
decidua, deciduum
Deciduous, as in *Larix decidua*

dioicus
dy-OY-kus
dioica, dioicum
With the male reproductive
organs on one plant and the
female on another, as in *Arunus
dioicus*

domesticus
doh-MESS-tih-kus
domestica, domesticum
Domesticated, as in *Malus
domestica*

The pear, *Pyrus communis*, is a commonly cultivated fruit tree with hundreds of cultivars grown commercially or in private collections.

funebris
fun-EE-bris
funebris, funebre
Connected to graveyards, as in
Cupressus funebris

generalis
jen-er-RAH-lis
generalis, generale
Normal, as in *Canna* × *generalis*

hortensis
hor-TEN-sis
hortensis, hortense
hortorum
hort-OR-rum
hortulanus
hor-tew-LAH-nus
hortulana, hortulanum
Relating to gardens, as in
Lysichiton × *hortensis*

incomptus
in-KOMP-tus
incompta, incomptum
Without adornment, as in
Verbena incompta

insititius
in-si-tih-TEE-us
insititia, insititium
Grafted, as in *Prunus insititia*

insulanus
in-su-LAH-nus
insulana, insulanum
insularis
in-soo-LAH-ris
insularis, insulare
Relating to an island, as in *Tilia
insularis*

lentus
LEN-tus
lenta, lentum
Tough but flexible, as in *Betula
lenta*

margaritaceus
mar-gar-ee-tuh-KEE-us
margaritacea,
margaritaceum
margaritus
mar-gar-EE-tus
margarita, margaritum
Relating to pearls, as in
Anaphalis margaritacea

mas
mas
masculus
MASK-yoo-lus
mascula, masculum
With masculine qualities, male,
as in *Cornus mas*

militaris
mil-ih-TAH-ris
militaris, militare
Relating to soldiers; like a
soldier, as in *Orchis militaris*

mutabilis
mew-TAH-bih-lis
mutabilis, mutabile
Changeable, particularly relating
to color, as in *Hibiscus mutabilis*

neglectus
nay-GLEK-tus
neglecta, neglectum
Previously neglected, as in
Muscari neglectum

❦ The name of the male
Cornelian cherry, *Cornus
mas*, probably relates to the
perception of physical strength
associated with masculinity –
the female Cornelian cherry,
C. sanguinea, has a less dense
woody structure.

 The bird of paradise, *Strelitzia reginae*, no doubt has many regal qualities but both its genus and species name relate to a particular queen, Charlotte, wife of George III and Duchess of Mecklenburg-Strelitz.

noli-tangere
NO-lee TAN-ger-ee
"Touch not" (because the seed pods burst), as in *Impatiens noli-tangere*

paradisi
par-ih-DEE-see
paradisiacus
par-ih-DEE-see-cus
paradisiaca, paradisiacum
From parks or gardens, as in *Citrus × paradisi*

paradoxus
par-uh-DOKS-us
paradoxa, paradoxum
Unexpected; paradoxical, as in *Acacia paradoxa*

pluvialis
ploo-VEE-uh-lis
pluvialis, pluviale
Relating to rain, as in *Dimorphotheca pluvialis*

poeticus
po-ET-ih-kus
poetica, poeticum
Relating to poets, as in *Narcissus poeticus*

redivivus
re-div-EE-vus
rediviva, redivivum
Revived; brought back to life (e.g. after drought) as in *Lunaria rediviva*

reginae
ree-JIN-ay-ee
Relating to a queen, as in *Strelitzia reginae*

religiosus
re-lij-ee-OH-sus
religiosa, religiosum
Relating to religious ceremonies; sacred, as in *Ficus religiosa*, under which the Buddha attained enlightenment

remotus
ree-MOH-tus
remota, remotum
Scattered, as in *Carex remota*

scabiosus
skab-ee-OH-sus
scabiosa, scabiosum
Scabrous; relating to scabies, as in *Centaurea scabiosa*

sclarea
SKLAR-ee-uh
From the Latin *clarus*, clear, as in *Salvia sclarea*

semperflorens
sem-per-FLOR-enz
Ever-blooming, as in *Grevillea × semperflorens*

sempervirens
sem-per-VY-renz
Evergreen, as in *Lonicera sempervirens*

sensibilis
sen-si-BIL-is
sensibilis, sensibile
sensitivus
sen-si-TEE-vus
sensitiva, sensitivum
Sensitive to light or touch, as in *Onoclea sensibilis*

sterilis
STER-ee-lis
sterilis, sterile
Infertile; sterile, as in *Potentilla sterilis*

subsessilis
sub-SES-sil-is
subsessilis, subsessile
Fixed, as in *Nepeta subsessilis*

tectorum
tek-TOR-um
Of house roofs, as in
Sempervivum tectorum

tenellus
ten-ELL-us
tenella, tenellum
Tender; delicate, as in *Prunus
tenella*

tener
TEN-er
tenera, tenerum
Slender; soft, as in *Adiantum
tenerum*

terminalis
term-in-AH-lis
terminalis, terminale
Ending, as in *Erica terminalis*

tremulus
TREM-yoo-lus
tremula, tremulum
Quivering; trembling, as in
Populus tremula

trivialis
tri-vee-AH-lis
trivialis, triviale
Common; ordinary; usual, as in
Rubus trivialis

turbinatus
turb-in-AH-tus
turbinata, turbinatum
Swirling around, as in *Aesculus
turbinata*

vagans
VAG-anz
Widely distributed, as in *Erica
vagans*

variabilis
var-ee-AH-bih-lis
variabilis, variabile

varians
var-ee-ANZ

variatus
var-ee-AH-tus
variata, variatum
Variable, as in *Eupatorium
variabile*

varius
VAH-ree-us
varia, varium
Diverse, as in *Calamagrostis
varia*

vegetus
veg-AH-tus
vegeta, vegetum
Vigorous, as in *Ulmus* × *vegeta*

verus
VER-us
vera, verum
True; standard; regular, as in
Aloe vera

vivax
VY-vaks
Long-lived, as in *Phyllostachys
vivax*

viviparus
vy-VIP-ar-us
vivipara, viviparum
Producing plantlets; self-
propagating, as in *Persicaria
vivipara*

vulgaris
vul-GAH-ris
vulgaris, vulgare

vulgatus
vul-GAIT-us
vulgata, vulgatum
Common, as in *Aquilegia
vulgaris*

◀ The planting of green roofs
is gaining in popularity, and
one name indicates plants that
may be used for the purpose.
Originally a superstitious
protection against storms,
hens and chicks, *Sempervivum
tectorum*, is still grown on roofs
to this day.

Other names as names

Sometimes, locally used colloquial names or names from antiquity become incorporated into the formal scientific epithets given to plants. For example, the scientific name of the Himalayan cedar, *Cedrus deodara*, is simply a Latinization of the Indian name for the tree. These names provide little useful information to gardeners but do add something to the rich language of plant nomenclature that all can share in and enjoy.

agnus-castus
AG-nus KAS-tus
From the Greek *agnos*, lamb, and *castus*, chaste, as in *Vitex agnus-castus*

alkekengi
al-KEK-en-jee
From the Arabic for bladder cherry, as in *Physalis alkekengi*

andrachne
an-DRAK-nee
andrachnoides
an-drak-NOY-deez
From the Greek *andrachne*, strawberry tree, as in *Arbutus* × *andrachnoides*

aria
AR-ee-a
From the Greek *aria*, probably whitebeam, as in *Sorbus aria*

azedarach
az-EE-duh-rak
From the Persian for noble tree, as in *Melia azedarach*

calamagrostis
ka-la-mo-GROSS-tis
From the Greek for reed-grass, as in *Stipa calamagrostis*

calamus
KAL-uh-mus
From the Greek for reed, as in *Acorus calamus*

carota
kar-OH-tuh
Latin for carrot, as in *Daucus carota*

cepa
KEP-uh
From the Latin for an onion, as in *Allium cepa*

cerasus
KER-uh-sus
Latin for cherry, as in *Prunus cerasus*

ceterach
KET-er-ak
Derived from an Arabic word applied to spleenworts (*Asplenium*), as in *Asplenium ceterach*

cyparissias
sy-pah-RIS-ee-as
Latin name for a kind of spurge, as in *Euphorbia cyparissias*

deodara
dee-oh-DAR-uh
From the Indian name for the deodar, as in *Cedrus deodara*

epipactis
ep-ih-PAK-tis
Greek name for a plant thought to curdle milk, as in *Hacquetia epipactis*

◖ The bladder cherry, *Physalis alkekengi*, is related to the Chinese lantern, *P. peruviana*, and, like it, all parts are poisonous except the fruit. Its name ultimately derives from a Greek word for a species of *Physalis*.

erinus
er-EE-nus
Greek name for a plant, probably basil, as in *Lobelia erinus*

laurocerasus
law-roh-KER-uh-sus
From the Latin *cerasus*, cherry, and laurel, *laurus*, as in *Prunus laurocerasus*

padus
PAD-us
Ancient Greek name for a kind of wild cherry, as in *Prunus padus*

pepo
PEP-oh
Latin for a large melon or pumpkin, as in *Cucurbita pepo*

ptarmica
TAR-mik-uh
ptarmica, ptarmicum
Ancient Greek name for a plant, probably sneezewort, that caused sneezing, as in *Achillea ptarmica*

pulegium
pul-ee-GEE-um
Latin for pennyroyal, reputed to be a flea-repellent, as in *Mentha pulegium*

quamash
KWA-mash
Nex Perce (Native American) word for *Camassia,* especially *C. quamash*

rhoeas
RE-as
From the Ancient Greek *rhoias*, as in *Papaver rhoeas*

 Perhaps derived from an Indo-European word for horn, *carota* is Latin for the carrot, *Daucus carota*. Originally white, orange carrots are thought to have been first introduced to the Netherlands in the 17th century.

ritro
RIH-tro
Probably from the Greek *rhytros*, globe thistle, as in *Echinops ritro*

robur
ROH-bur
From the Latin for oak wood, as in *Quercus robur*

sasanqua
suh-SAN-kwuh
From the Japanese name for *Camellia sasanqua*

schafta
SHAF-tuh
From the Caspian vernacular name for *Silene schafta*

serpyllum
ser-PIE-lum
From the Greek word for a kind of thyme, as in *Thymus serpyllum*

siliquastrum

sil-ee-KWAS-trum

From the Latin for a plant with
pods, as in *Cercis siliquastrum*

spicant

SPIK-ant

Word of uncertain origin;
possibly a corruption
of *spica*, spike, tuft, as in
Blechnum spicant

stoechas

STOW-kas

From the Greek *stoichas*, rows, as
in *Lavandula stoechas*

strobus

STROH-bus

From the Greek *strobos*, a
whirling motion (cf. Greek
strobilos, pine cone), or the Latin
strobus, an incense-bearing tree
in Pliny, as in *Pinus strobus*

tangelo

TAN-jel-oh

From tangerine, *Citrus reticula*,
and pomelo, *C. maxima*, as in
Citrus × tangelo

The Latin for oak wood,
robur, is also the specific
epithet of the English oak,
Quercus robur, a valuable
source of food for wildlife.

tobira

TOH-bir-uh

From the Japanese name for this
shrub, as in *Pittosporum tobira*

totara

toh-TAR-uh

From the Maori name for this
tree, as in *Podocarpus totara*

trichomanes

try-KOH-man-ees

Relating to a Greek name for
fern, as in *Asplenium trichomanes*

tupa

TOO-pa

Local Chilean name for
Lobelia tupa

Index

arborescens	26	aubrietioides	87	berthelotii	118
arboreus	26	aucuparius	124	betaceus	87
arboricola	114	aurantiacus	16	betonicifolius	95
arbutifolius	**95**	aurantius	16	betulinus	87
archangelica	118	auratus	16	betuloides	87
arenarius	114	aureosulcatus	16	bicolor	20
arendsii	118	aureus	16	bicornis	64
arenicola	114	auricomus	16	bicornutus	64
arenosus	114	auriculatus	82	biennis	136
argentatus	12	auriculus	82	bifidus	64
argenteus	12	auritus	82	biflorus	42
argutifolius	48	australis	112	**bifolius**	48, **49**
argyrophyllus	12	austriacus	102	bifurcatus	26
aria	142	autumnalis	136	bignonioides	87
aristatus	64	avellanus	102	bilobatus	48
armandii	118	avium	124	bilobus	48
armatus	**58**	axillaris	64	bipinnatus	48
armeniacus	106	azedarach	142	biternatus	48
aromaticus	62	azoricus	102	**blandus**	**128**
artemisioides	86	azureus	16	blepharophyllus	82
articulatus	56			bodinieri	118
arundinaceus	86			bodnantense	102
arvensis	115			bonariensis	110
ascendens	26			borealis	112
asclepiadeus	87			botryoides	82
asparagoides	87			bowdenii	118
asper	58	babylonicus	106	brachybotrys	42
asperatus	58	baccans	54	brachycerus	64
asperrimus	58	bacciferus	54	brachyphyllus	49
asphodeloides	87	baldschuanicus	106	**bracteatus**	**42**
assa-foetida	62	balsameus	62	bracteosus	42
asteroides	87	bambusoides	87	bractescens	42
astilboides	87	banksianus	118	brevifolius	49
atlanticus	112	banksii	118	brevipedunculatus	42
atriplicifolius	95	bannaticus	102	brevis	34
atrocarpus	20	barbarus	138	bromoides	87
atropurpureus	20	barbinervis	58	bronchialis	126
atrorubens	20	basilicus	128	brunneus	20
atrosanguineus	20	beesianus	102	bryoides	87
atrovirens	20	bellidiformis	87	bufonius	124
attenuatus	64	berolinensis	102	bulbiferus	56

B

D

fastigiatus	28	floccigerus	58	fuscus	21
fastuosus	132	floccosus	58	futilis	132
fatuus	132	florentinus	103		
febrifugus	126	floribundus	44		
fecundus	129	floridus	44		
fenestralis	79	flos	44		

G

ferrugineus	21	fluitans	115		
fertilis	129	fluminensis	111		
festalis	129	fluvialis	115	galeatus	79
festivus	129	fluviatilis	115	galericulatus	79
fibrillosus	58	foeniculaceus	89	gallicus	103
fibrosus	58	foetidissimus	62	garganicus	103
ficoides	89	foetidus	63	gemmatus	79
ficoideus	89	foliolosus	50	gemmiferus	44
filamentosus	66	foliolotus	50	generalis	139
filarius	66	**foliosus**	**50**	genistifolius	96
filicifolius	96	fontanus	115	geoides	89
filipendulus	89	formosanus	107	gibbosus	66
filipes	56	formosus	129	gibbus	66
fimbriatus	66	forrestianus	119	giganteus	34
firmatus	129	forrestii	119	gilvus	14
firmus	129	fortunei	119	glabellus	59
fissilis	**66**	foveolatus	66	**glaber**	**59**
fissuratus	66	fragarioides	89	glabratus	59
fissus	66	fragilis	56	glabrescens	59
fistulosus	56	fragrans	63	glabriusculus	59
flabellatus	44	fragrantissimus	63	glacialis	115
flaccidus	66	fraxinifolius	96	gladiatus	79
flagellaris	79	frigidus	115	**glanduliferus**	**44**
flagelliformis	79	frutescens	28	glaucescens	14
flammeus	17	fruticans	28	**glaucophyllus**	**14**
flavens	13	fruticosus	28	glaucus	14
flaveolus	13	fucatus	17	globiferus	66
flavescens	13	fugax	66	globosus	37
flavicomus	13	fulgens	66	globularis	37
flavidus	13	fulgidus	66	globuliferus	66
flavovirens	13	fulvidus	14	glomeratus	44
flavus	14	fulvus	14	gloriosus	129
flexicaulis	56	funebris	139	glutinosus	66
flexilis	56	furcans	28	gongylodes	37
flexuosus	56	furcatus	28	gossypinus	89

H

I

M

N

O

P

Q

R

S

T

tabularis	38
tabuliformis	38
takesimanus	109
taliensis	109
tanacetifolius	98
tangelo	144
tanguticus	109
tardiflorus	137
tardus	137
tasmanicus	113
tatsienensis	109
tauricus	109
taxifolius	**98**
tazetta	80
tectorum	**141**
tenax	61
tenellus	141
tener	141
tenuicaulis	57
tenuifolius	53
tenuis	30
tenuissimus	30
tequilana	111
terebinthifolius	98
teres	30
terminalis	141
ternatus	53
terrestris	117
tessellatus	70
testaceus	18
testudinarius	76
tetragonus	70
tetrandrus	47
tetrapterus	**55**
textilis	127
thalictroides	93

thibetanus	109
thibeticus	109
thunbergii	120
thymifolius	98
thymoides	93
thyrsoides	80
thyrsoideus	80
tiarelloides	93
tigrinus	76
tinctorius	127
tingitanus	113
titanus	**35**
tobira	144
tomentosus	61
tommasinianus	121
torreyanus	121
tortilis	30
tortuosus	30
tortus	30
totara	144
toxicarius	133
trachyspermus	55
tragophylla	76
transitorius	137
traversii	121
tremulus	141
tri-	70
triacanthos	61
triandrus	47
triangularis	70
triangulatus	70
tricho-	70
trichomanes	144
trichotomus	57
tricolor	22
tricuspidatus	**53**
trifasciata	71
trifidus	71
triflorus	47
trifoliatus	53
trifolius	53

trifurcatus	71
trigonophyllus	53
trimestris	**137**
trinervis	53
tripartitus	71
triplinervis	53
tripteris	71
tripterus	71
tristis	133
triternatus	53
trivialis	141
truncatus	71
tschonoskii	121
tuberosus	**30**
tubiferus	38
tubulosus	38
tulipiferus	93
tuolumnensis	111
tupa	144
turbinatus	141
turczaninowii	121
typhinus	93

U

uliginosus	117
ulmaria	93
umbellatus	47
umbrosus	117
uncinatus	71
undatus	53
undulatus	53
unedo	63
unguicularis	76
unguiculatus	76
uni-	71
uplandicus	104
urbanus	117

V

X

Y

Z

Credits

All of the illustrations featured in this book are either © Royal Horticultural Society or are in the public domain.